*Fallacy of the Green Movement
and Climate Change*

Fallacy of the Green Movement and Climate Change

Personal Collection of Papers and Responses Disputing Positions of the EPA-Environmental Protection Agency, "An Inconvenient Truth" by Al Gore, and Responses to Media Articles

Joseph E. Schramek

Copyright © 2015 by Joseph E. Schramek.

ISBN: Hardcover   978-1-5035-7337-6
       Softcover   978-1-5035-6739-9
       eBook   978-1-5035-6738-2

All rights reserved. No part of this book may be reproduced or transmitted in any form or by any means, electronic or mechanical, including photocopying, recording, or by any information storage and retrieval system, without permission in writing from the copyright owner.

Scripture quotations marked HCSB are from the Holman Christian Standard Bible®. HCSB®. Copyright ©1999, 2000, 2002, 2003, 2009 by Holman Bible Publishers. Used by permission. Holman Christian Standard Bible®, Holman CSB®, and HCSB® are federally registered trademarks of Holman Bible Publishers

Any people depicted in stock imagery provided by Thinkstock are models, and such images are being used for illustrative purposes only.
Certain stock imagery © Thinkstock.

Print information available on the last page.

Rev. date: 06/24/2019

To order additional copies of this book, contact:
Xlibris
1-888-795-4274
www.Xlibris.com
Orders@Xlibris.com
710777

CONTENTS

Topic	Page
Author's Note	xi
The Green and Carbon Footprint Movements in the World-Why?	1
EPA's Position That Carbon Dioxide-CO_2 is a Pollutant, Is Dead Wrong	6
Acid Rain-Sulfer Dioxide (SO_2) and Nitrogen Oxide (NO_2) Products from Dirty Coal Burning Process	10
• Attachment I	19
• Attachment II Sulfuric and Sulfurous Acids Produced from Sulfur Dioxide-SO_2	20
• Attachment III Sources of Nitrogen Dioxide NO_2 and Nitric Acid Coal Fired Electrical Plants	24
• Attachment IV Effects of Acid Rain - Forests	27
Convenient Lies-The Measurement of Carbon Dioxide at Volcano Sights	29
Letter to President Lou Anna K. Simon of Michigan State University, "Methane Gas Versus Coal-Powered Power Plants",	40
Responses to Various Claims of the Environmental Movement	43
VW 590,000 2009-2016 Vehicles Equipped With 2 L and 3-L Diesel Engines with Defeat Devices-To Cheat on EPA's NO_x Emissions Requirements	51
"Water-A Deeper Understanding of Our Most Vital Resource" Article in the Fall 2017 MSU Alumni Magazine	59

Flint Water Contamination from Switching from Detroit Water Supply To Water Originating in the Flint River	63
Letters to the Editor of the Dearborn Press and Guide-May 7, 2011 "Global Warming-Real or Imagined"	76
Letter To the Editor of the Detroit News-May 4, 2011	79
MSU Gets Wet-Letter to Editor of the MSU Alumni Magazine	82
Response to "earth is dying" and "Carbon Diet and Footprint" Advocates	83
Letter to Nolan Finley on Detroit News Article-February 4, 2014 Faith leaders must speak out on climate change-Charles Morris of Madonna Univ	86
Response to Hackett High School on New Solar Panels-Goes Greener	89
Response to Tim Powers Editor of the Dearborn Press and Guide	93
Gov. Granholm's State Address-Coal Power Plants Spew Dirty Carbon	95
Letter to Detroit News-Jim Nash's Letter, "The Problem with Fracking"	98
Detroit News Editors-"Michigan needs a 'no-regrets' strategy"	100
Letter to Detroit News-Artic Study Warns of Melt and 5 Foot Rise in Ocean	102
Letter to MSNBC-Chris Jansing's Show Today-Gore's 20 Feet Rise in Ocean	105
Responses to Recent Articles in Detroit News October, 2014 "Buyers slow to connect with electric vehicles", and "GM Chief: Time to Get Candid"	107
Response to Detroit News Article on January 1, 2015-"Recall woes follow GM into New Year"	110
Dearborn Press and Guide Article Published February 26, 2012- Your Point of View-"Support for new Goodwill store at West Dearborn Location	115

Detroit News-HMO Versus BCBS Costs-March 6, 1979	118
"A Christmas Favor", By My Mother, Gertrude Schramek	120

Appendix

- Top Points and Arguments Used to Validate "Fallacy of the Green Movement and Climate Change" Claims from the 1st Four Chapters of the Book — 124
- Author's Statement at the Goodread's in Support of My Book — 134
- Author's Rebuttal to Glen R. Stott's Criticism of My Book As Recorded in Goodread's September 21, 2015 — 138
- Author's Rebuttal to Glen R. Stott's Criticism As Shown in His E-Mail Back to Schramek-August 31, 2018 — 148
- Lead Contents in Food-Meats, Fish, Drinks, and Vegetables — 166

Epilogue — 173

To Those Who Seek the Truth to Become More Knowledgeable About the Fallacy of Climate Change and the Importance of Carbon Dioxide for the Continuance of Human and Vegetation Life on Earth

Joe Schramek-Class of 1957

Joe Schramek-Current Photo

Author's Note

My interest in the composition of the air around us started in High School, and College, and all of my life. As an 8th grader, my first troubling experience with weather occurred on June 8, 1953 at about 8:30 PM at my family's home in Lansing, Michigan. Their home was along the Grand River across the river from the Oldsmobile Factory. That night my parents were holding a card party for friends from our church, St. Casimir's. At the front door, the sky was blackened and intense continuing lighting flashes began and lasted about 30 minutes. I never saw before such a scary weather event. The next day, it was news across the country that at 8:30 PM on June 8th, a section of Flint, Michigan called Beecher had similar intense lightning flashes before one of the top ten worst tornadoes (F5) hit them and killed 116 people. This impacted me greatly, it made me wonder how tornado's form and the causes of this phenomenon.

In 1956, my Dad took a new job in Kalamazoo, and I attended St. Augustine High School there for my last 4 months of my Junior and all of my Senior Year of High School. In high school, I took Pre-College science courses of Math,

Chemistry and Physics. For college, I attended Western Michigan University on a Pre-Engineering Program that included intensive Math, Chemistry and Physics courses. As a student I did quite well as I was the 6th Best Freshman Chemistry student and a Honor's Calculus Student. As a Junior, I transferred to Michigan State University where I selected the Electrical Engineering curriculum. My Dad was a Mechanical Engineering Graduate in 1925 from MSU. I didn't want to do what my dad did so I took Electrical Engineering as I was good at Applied Math. It's hard to shake inherited whims and talents of my mother and father, and after graduation I was hired by Ford Motor Company in Chassis Engineering. Initially I was put on a Grad-Trainee Rotational Program for 2 years, where I worked in about 8 different job areas. My last 3 month stint was in the Chassis Frame Group. I found the work there interesting and challenging and they liked my work effort and performance. I joined the Chassis Frame Group took over the testing and laboratory prove out of the new 1965 Ford Torque Box Frame. It was a successful effort in that the frame was used on all Ford large body frame vehicles until 2011 when the Ford Crown Victoria Police vehicle production was stopped. I was promoted to Supervisory positions and continued with a successful career there for 44 years with a retirement in 2006.

Because of my knowledge of elementary electrical circuit theory, chemistry, and physics, as early as high school I started to question anything pertaining to atmosphere and it's effect on the flight length and control of golf balls, speed and curve ball trajectories of pitched baseballs, distance of baseball hits, and distance required for takeoff of airplanes for various atmospheric and temperature conditions. Very early in my life, I probably was one of a few people who accurately call hot humid air as ***light air*** in place of the common response of individuals who call it ***heavy air***. Hot humid air feels heavy, because the lighter water vapor molecules replace the heavier oxygen and nitrogen molecules and with less available oxygen molecules in the air, it is harder to breath for most people.

While at Ford Motor Company, the Government's EPA Agency began rule-making on Automotive Emissions and MPG-Miles Per Gallon for vehicles beginning on Mid- 1970 vehicles. This was accomplished with catalytic convertors that drastically eliminated incomplete combustion of gas and limited amounts of undesirable carbon and nitrous oxides. In my paper's you will learn about how much liquid water and carbon dioxide weight is created by burning a gallon of gasoline. You will be amazed! Also, you will be amazed where all the carbon dioxide goes and why it is so necessary for continued life here on earth.

The Green and Carbon Footprint Movements in the World-Why?

The Fallacy of the Movements

The Green and Carbon Foot Print Movements across the world today are associated with environmental scientists and interest groups who feel the increasing use of fossil fuels (oil, gas, and coal) by man leads to higher ambient atmospheric temperatures. As a result, they feel the temperature rise will cause more climatic disasters, and melting of the polar ice caps causing a rise in ocean levels which would reduce the land available for life across the continental seashores of the earth. Over the past 10-15 years, none of these predictions have come to fruition or proven as a causal factor. Every time a major environmental disaster, flooding, hurricane, tornado, or arid condition occurs, activists blame it on the increase in fossil fuel consumption and burning by man.

Joseph E. Schramek

Why burning of fossil fuels has no effect on the atmospheric temperatures and any of the climatic related disasters?

Burning of coal and gasoline produces water vapor and carbon dioxide. Both products are greenhouse gases. Water vapor is the more potent of the two greenhouse gases. Water vapor can reach a high of 4% of the atmosphere over the ocean, while carbon dioxide is only .0392% of the atmosphere with an annual range of +.0003% in winter months and -.0012% at the end of the growing season. The water vapor molecule has a lighter molecular weight (18) but higher amounts water vapor in the atmosphere (up to 40,000 ppm (parts per million)) for water vapor versus a relatively constant 392 ppm for carbon dioxide-molecular weight (44), and has a much larger impact (100 times) on the earth's atmospheric pressure. Since the amount of carbon dioxide is almost constant at 392 ppm, and the water vapor molecule varies from 0-4% of the atmosphere, water vapor molecule is totally responsible for our violent hurricanes and tornadoes and storm damages from high straight line winds in storms.

Why is the water vapor molecule responsible for all climatic disasters?

In high pressure systems (Barometric reading of 29-30 inches of Mercury), we experience good weather with few clouds and fresh air with the highest percentage of oxygen-21% and lowest humidity in the air for comfortable breathing. In low atmospheric pressure systems (Barometric reading of 26-28 inches of Mercury) the air is very warm and humid-3% water vapor and with reduced oxygen-about 20% in the atmosphere for poorer breathing conditions and discomfort as perspiration condenses on skin without evaporation and cooling of the skin on our body. Low pressure systems are sometimes incorrectly labeled heavy air. The low pressure systems provide the water molecules which produce rain when the system is confronted with a colder temperature high pressure front that generally comes from the Northwest in the Northern Hemisphere. These fronts produce the most devastating tornadoes and thunderstorms with high straight line winds. For some reason, the scientists ignore the water vapor molecule as the cause for all the climatic temperature disasters and global warming (real or unreal or insignificant). Because there is increasing burning of fossil fuels in the world, they incorrectly choose the other greenhouse gas, carbon dioxide as the culprit, even though the percentage of carbon dioxide in the atmosphere is constantly in the range of 392 ppm or .0392%.

What keeps the percentage of carbon dioxide relatively constant at 392 ppm or .0392%?

What keeps the percentage of carbon dioxide relatively constant at 392 ppm or .0392% is that plant life has a voracious appetite for the carbon dioxide and water molecule to create vegetation to cover the earth's soil and provide food supplies for animal and human life.

A simple analysis is that the voracious appetite of plant life for carbon dioxide is matched almost equally to the creation of carbon dioxide through burning hydrocarbons for heating and energy for manufacturing processes.

Why is the burning of fossil fuels and hydrocarbons important for continuation of life here on earth?

Hydrocarbons in nature include all plant and animal life, and fossil fuels (oil, and coal). Hydrocarbons are organic compounds that are produced naturally from water and carbon dioxide through photosynthesis. Over eons of years, the trapping and decaying of these remnants through seismic activity, transformation and relocation of the earth's continental plates and oceans results in the finding them deep in the earth as fossil fuels; Including: coal, and oil. Combustion of coal and oil products (gasoline, diesel fuel, and other petroleum variations) for heating and energy sources by burning, and digestion of food products by man and animal life results in re-creation of the water vapor molecule and carbon dioxide molecule. As an example, an automobile burning one gallon of gas will produce 1 gallon of liquid water when the water vapor is condensed into water. It also produces copious amounts of carbon dioxide molecules that produces and maintains our Oxygen supply for human and animal life. Without Oxygen, we could not ignite fossil fuels and waste to produce energy for survival here on earth. Any efforts to reduce the percentage of carbon dioxide in the atmosphere could eventually reduce the earth's available vegetation for animal food supplies.

Why is it important to maintain the available Oxygen levels in our atmosphere at 21%?

The most devastating effect of the Green Movement would be the loss or reduction of the Oxygen molecule levels in the atmosphere from a rather constant level of 21% to lower levels. High humidity reduces the Oxygen molecule count in the atmosphere as the water molecule causes the atmosphere pressure dropping from 30 inches of Mercury to 26-27 inches for near hurricane and tornado weather conditions. Some have difficulty with breathing under these conditions especially for seniors and any workers doing heavy labor. High humidity prevents workers from evaporating their body sweat to cool their body and blood and in some cases heat strokes may occur. Oxygen molecules are only added to the atmosphere from plant life and vegetation during the process of photosynthesis. If the current Green and Carbon Footprint Movements ever reach its desired goals for significantly reducing the carbon dioxide molecule percentage of .0392% or 392 ppm levels, humans and all animal life would begin to experience breathing issues causing many variations of health concerns including possible death. The Oxygen molecule in nature only comes from the Carbon Dioxide molecule through the photosynthesis process that produces our vegetation and plant life on earth.

EPA's Position That Carbon Dioxide-CO_2 is a Pollutant, Is Dead Wrong

What some of the uses for the Pollutant, Carbon Dioxide?

Carbon Dioxide has many uses for humans. Carbon Dioxide is bubbled into water to form carbonic acid. Carbonic acid is a weak acid, and the carbon dioxide bubbles into the air from the drink or when agitated during pouring. Carbonic acid is added into beverages (sodas, and beer) to improve taste and freshness. Frozen Carbon Dioxide is known as dry ice and is used to keep medicines, and refreshments refrigerated for long periods of time when electrical power is lost or products are being shipped by mail. The main use for this pollutant is the prime ingredient in the production of oxygen gas for the atmosphere though the process of photosynthesis as described below:

Photosynthesis is the process whereby plants using light energy from the sun to convert carbon and water pollution in the soil to produce glucose sugar and oxygen gas through a series of reactions. Water and Carbon Dioxide gases in the atmosphere are the main greenhouse gases in the atmosphere. The EPA (Environmental Protection Agency) has declared that Carbon Dioxide is a pollutant because it causes unfavorable climate change. Water vapor or humidity is the most prevalent greenhouse gas at a maximum of 4% of the atmosphere at sea level and carbon dioxide pollution is relatively constant at .0392% or 100 times lesser than gaseous water vapor in the atmosphere. The chemical equation for photosynthesis is shown below:

$$\text{Carbon Dioxide} + \text{Water} \xrightarrow[\text{Chlorophyll}]{\text{Light}} \text{Glucose} + \text{Oxygen}$$
$$6CO_2 \qquad 6H_2O \qquad\qquad\qquad C_6H_{12}O_6 + 6O_2$$

In producing vegetation in the form of farm crops, forests, all green plant life, the glucose from this process is further processed to produce carbohydrates, the storage mechanism of chemical energy. The oxygen gas produced from the photosynthesis process is the sole source of new oxygen in the atmosphere for all human and animal life on earth. Currently, because of the voracious appetite of plant life for the pollutant carbon dioxide, oxygen gas remains at a relative 21% of the total atmosphere on earth, thanks to the natural environmental process of photosynthesis. Most carbohydrates have a ratio of 1:2:1 of carbon, hydrogen, and oxygen, respectively, or by weight, Oxygen 50%, Carbon 44% and Hydrogen 6%. The conclusion is that the vegetation on earth consumes all of the pollutant, Carbon Dioxide.

Joseph E. Schramek

How is Carbon Dioxide formed here on earth?

Carbon dioxide is produced by combustion of coal, and petroleum products and all hydrocarbons, the fermentation of sugars in beer and winemaking and by respiration of all living organisms. It is exhaled in the breath of humans and animals. It is emitted from volcanoes, hot springs, geysers and other places where the earth's crust is thin and is freed from carbonate rocks by dissolution. CO_2 is also found in lakes, at depth under the sea and commingled with oil and gas deposits.

How many Carbon, Oxygen, and Hydrogen Atoms exist on Earth?

With the exception of Hydrogen, all the carbon and oxygen atoms that existed here on earth when it settled in orbit around the sun are still here today. Hydrogen is the only element that can achieve an escape velocity greater that the earth's gravitation force. Some NASA space missions removed a minimal number of atoms of all of the elements of the earth when they left the earth's gravitational forces. Any escaped hydrogen gas that left earth most likely ended up in the surface of the Sun eventually.

The atmosphere contains the greenhouse gases in varying amounts from: water vapor at a maximum level of 4% at sea level, and carbon dioxide at a relatively constant level of .0392% with an annual range of +.0003% in winter months and -.0012% at the end of the growing season depending on the seasons in the Northern and Southern

Hemispheres of the world. Water vapor or humidity can lower atmospheric pressures from 30 mm for low humidity and fair weather and 26 mm for high humidity for tornado and hurricane weather conditions. Atmospheric conditions are largely dependent on High and Low Pressure Systems. High Pressure Systems at 30 mm have clear blue skies and low humidity while Low Pressure System at 28-26 mm have high humidity and temperature conditions and gray threatening skies. Because Carbon Dioxide stays at insignificant levels in the atmosphere, it does not have any effect on weather conditions despite the EPA ruling it a pollutant.

Acid Rain-Sulfer Dioxide (SO_2) and Nitrogen Oxide (NO_2) Products from Dirty Coal Burning Process

What is Acid Rain?

Acid rain has a PH of 5.5 which is a weak acid. Others may argue whether the acid is predominately HCO_3 Carbonic Acid or others including: nitric, sulfurous, and sulfuric acids too. The emissions providing the precursors of the acid rain formation come from both natural sources, such as volcanoes, and decaying vegetation, and man-made sources, primarily emissions of sulfur dioxide (SO_2) and nitrogen oxides (NO_X) resulting from combustion of fuel and coal. When these gases are released from their sources, prevailing winds blow these gases and compounds across state and national borders. These winds dilute the concentration of the gases such that none have concentrations over .00002% in total with the exception of CO_2 at .0004% of the atmosphere.

Why is this definition of Acid Rain incorrect?

The definition is incorrect based on the following:

H_2SO_4-Sulfuric Acid

- Sulfuric acid, H_2SO_4, cannot be created from SO_2 in the atmosphere. The EPA erroneously reports that H_2SO_4 Sulfuric Acid can be produced in the air from SO_3, but it requires the presence of the catalyst, NO_2 Nitrogen Dioxide. This has no chance or the remotest of possibilities of occurring in the atmosphere as both SO_3, and NO_2 gases are present at only .00005% of the air. (Attachment I and II)
- SO_2 Sulfur Dioxide, occurs in the atmosphere at 1 part/billion.
- Sulfuric acid, H_2SO_4, is manufactured by combining SO_2 with Oxygen to produce SO_3 using a catalyst of Platinum or Vanadium Pentoxide with a temperature of 450 degrees Celsius and pressure of 1-2 atmospheres. Then, SO_3 Sulfur Trioxide, is added to H_2SO_4 Sulfuric Add to form oleum, $H_2S_2O_7$ Oleum is added to water to produce a concentrated level of Sulfuric Acid. This is the method to produce Sulfuric Acid in volume and a high concentration of the acid. In the atmosphere, SO_2, at 1 part/billion must have a catalyst, and high temperature, and find a catalyst, NO_2 Nitrogen Dioxide that exists in the atmosphere at .00002%, as EPA suggests exists, and then be subjected to complete the transition to SO_3 Sulfur Trioxide and unites with water vapor to form a very diluted weak Sulfuric Acid concentration.

Joseph E. Schramek

- Sulfuric acid is used in large quantities by the iron and steelmaking industry to remove oxidation, rust and scaling from rolled sheet and billets prior to sale to the automobile and major appliances industry.
- Aluem H_2SO_7 is made from sulfuric acid. It is a dehydrator and whitening agent for the manufacture of paper. Another important use for sulfuric acid is for the manufacture of aluminum sulfate, also known as paper maker's alum.
- Sulfuric acid has a wide range of applications including domestic acidic drain cleaner, electrolyte in lead-acid batteries in automobiles and trucks and various cleaning agents. It is also a central substance in the chemical industry. Principal uses include mineral processing, fertilizer manufacturing, oil refining, wastewater processing, and chemical synthesis. It is widely produced with different methods, such as contact process, wet sulfuric acid process and some other methods.
- Should any H_2SO_4, Sulfuric Acid be created in the atmosphere, it is hygroscopic (readily removes water from the atmosphere) and becoming more dilute and weaker. Also, it is a very strong dehydrator and oxidizing agent and it will attack any metal, fabric, vegetation, and flesh it comes in contact with. Because of its weak concentration, it would be difficult to notice any change, if it truly exists in the atmosphere.
- Attachment I and II shows no appreciable SO_2 concentration in the dry atmosphere for elements and gaseous compositions, as it readily makes the H_2SO_3 Sulfurous Acid molecule in the gaseous state in the atmosphere.

H_2SO_3 Sulfurous Acid

- Sulfurous acid, H_2SO_3, is simply produced by combining the gas SO_2 with water vapor molecule to form a gaseous molecule of H_2SO_3. H_2SO_3 only exists as a gas molecule and does not exist on record as a liquid solution. It reverts back to SO_2 Sulfur Dioxide gas in a liquid state.

HNO_3 Nitric Acid

- NO_2 Nitrogen Dioxide occurs in the dry atmosphere at .00002%. When the gas molecule is mixed in the atmosphere with water vapor the water vapor molecule, it forms one molecule of HNO_3 Nitric Acid. The acid molecule created is very sparse in the atmosphere. Its life is short because it is a strong oxidizer and oxidizes metals and vegetation. Its damage would be insignificant because of the weakness and strength of the acid overall. (Attachment III)
- Nitrous oxide (N_2O) occurs in the dry atmosphere at .0000325%. It does not produce Nitric Acid in the atmosphere or anywhere else.

Why does the EPA require Coal Burning Power Plants Meet Emission Requirements for SO_2 and NO_x, When Acid Rain Does Not Kill Trees?

EPA's Assessment of Acid Rain on Forest Degradation

From Attachment IV, For years, some concern exists for the vegetation, and forest trees from Maine to Georgia in the

Joseph E. Schramek

Shenandoah and Great Smoky Mountain National Parks at the higher elevations over staying healthy thru the growing season. Some believe that the cause of this demise may be caused by acidic gases from coal-fired power plants and environmental stressors such as: insects, disease, drought, or very cold weather. Future work to understand the issue pertains to these areas:

Acid Rain on the Forest Floor

It is known that rain water has a ph of 5.5 which is a weak acid. Some may argue whether the acid is predominately H_2CO_3 Carbonic Acid or others. Carbonic acid is a buffer that tends to keep the acidity from increasing. The ability of forest soils to resist, or buffer, acidity depends on the thickness and composition of the soil, as well as the type of bedrock beneath the forest floor.

How Acid Rain Harms Trees

Studies indicate that acid rain does not kill trees. Efforts seem to be centered on finding some relationship to coal-fired electric power plant emissions as a chief cause for the conditions. Others want to blame runoff of rain water for washing away key nutrients of the soil and exposure of toxic aluminum as a result of acid rain. Actually, the runoff of rain water will remove soil and nutrients necessary for growth and health of the tree. Overlooked is the need of rain water in the hot summer as a condition causing the continuation of the photosynthesis process where water and carbon dioxide provide the key elements for vegetational growth in the trees and other plant life affected.

Is the EPA knowledgeable or truthful on why some trees died at high elevation at the Eastern United States due to SO_2 and NO_x gases that may or may not have been transformed into Sulfuric and Nitric Acids?

EPA admitted above that Acid Rain cannot kill trees alone. Then the EPA doesn't know, and is guessing that Acid Rain and along with other causes resulted in the death of trees in the forests areas located at high elevations in the Eastern United States. Since high elevation occurs at these sites on the Eastern half of the United States with highly populated cities occurring on both sides of these mountain ranges, and these areas have Coal Fired Power Plants close by to provide the electrical energy to support the large industrial cities and residences that border these mountain ranges.

With regard to the other possible causes of trees dying, most mountain areas have a natural forest line where trees exist and trees are lacking due to the extreme lower temperatures at higher altitudes. Other causes are droughts-lack rain water supply, lack of soil to support root structures, over population of trees-lack of forest management, diseases, insects, and wind damage. As an examples, Elm trees are gone to Dutch Elm disease, and Ash Trees are gone due the Emerald Ash Borer.

The EPA uses the stronger levels (PH's of 2.4) of acids in the lakes of Eastern US areas in defense of its requirements for SO_2 and NO_X emissions, when "Acid rain" is a popular term referring to the deposition of wet (rain, snow, sleet, fog, cloudwater, and dew) and dry (acidifying particles and gases) acidic components. Distilled water, once carbon dioxide is removed, has a neutral pH of 7. Liquids with a pH less than 7 are acidic, and those with a pH greater than 7 are alkaline. "Clean" or unpolluted rain has an acidic pH, but usually no lower than 5.7, because carbon dioxide and water in the air react together to form carbonic acid, a weak acid according to the following reaction:

$$H_2O + CO_2 = H_2CO_3$$

Again, going from a PH of 5.7 to 2.4 is an extreme value in some isolated cases. More study is needed to determine why the higher acidic values occur. Simply, having a drought season would increase the acidity of the lake. EPA uses the presence of an acid with Carbon Dioxide-the pollutant as declared by the non-Chemists of the United States Supreme Court, even though, it is our only source of Oxygen supply for all human life and vegetation on Earth.

How much tailpipe Carbon Dioxide (CO_2) and Water (H_2O) is created from burning one gallon of fuel?

The amount of CO_2 created from burning one gallon of fuel depends on the amount of carbon in the fuel. After combustion, a majority of the carbon is emitted as CO_2 and very small amounts as hydrocarbons and carbon monoxide.

Carbon content varies by fuel, and some variation within each type of fuel is normal. The EPA and other agencies use the following average carbon content values to estimate CO_2 emissions:

- CO_2 Emissions from a gallon of gasoline: 8,887 grams or 19.6 lbs. of CO_2 / gallon1
- CO_2 Emissions from a gallon of diesel: 10,180 grams or 22.4 lbs. of CO_2 / gallon$_2$
- Water Vapor Emissions from a gallon of gasoline: 3,780 grams or 8.34 lbs. of H_2O / gallon.

Vehicles that use diesel fuel generally have higher fuel economy than comparable gasoline vehicles. However, when comparing carbon dioxide emissions, the higher CO_2 emissions from diesel fuel partially offset the fuel economy benefit.

The irony of the amounts of Carbon Dioxide and Water produced from burning coal and gasoline is that the EPA never mentions that Carbon Dioxide is the miracle gas that produces all of our food through photosynthesis along with the water created in burning of hydrocarbons, and the only oxygen supply used by mankind on earth. They only see them as pollutants and detrimental to existence of mankind on earth, unbelievable.

Joseph E. Schramek

Why does the EPA require Electric power stations that burn coal be equipped with flue gas desulfurization scrubber processes to reduce SO_2 by 95% in the flu gas while producing large quantities of gypsum?

EPA's position is that SO_2 from the Power Plant coal burning process results in Sulfuric Acid H_2SO_4. This paper casts significant doubt that their position is not true, and that H_2SO_4 Sulfuric Acid is produced in the atmosphere and result in acid rain from SO_2 gas emitted by the flue gases of the Coal Fired Power Plants.

The fact that gypsum (Plaster of Paris) used to make dry wall for construction of buildings and homes is created in large supplies by the wet flue gas desulfurization scrubber process used at most Eastern United States Power Plants is meaningless, since the unique conditions of the White Sands National Monument in the US state of New Mexico have created a 710 km² (270 sq mi) expanse of white gypsum sand, enough to supply the construction industry with drywall for 1,000 years.

Attachment I

The atmosphere is largely composed of Nitrogen, Oxygen, and Argon. Water or water vapor averages .25% of the atmosphere, but it can raise to 4-5% of the atmosphere in drastic low pressure systems that are the chief source of hurricanes, tornadoes, cyclones, straight-line high winds, and excessive wind and flood damages as a result. Water vapor is the largest concentration of the greenhouse gases and can be 100 times more that carbon dioxide in high temperature/high humidity conditions of an extreme low pressure system involving 26-27 barometric readings. The remaining gases on the list are trace gases, including: greenhouse gases: carbon dioxide, methane, nitrous oxide, and ozone, and various natural, chemical, and industrial gases.

Composition of dry atmosphere,

Gas	Volume
Nitrogen (N_2)	78.%
Oxygen (O_2)	21%
Argon (Ar)	.9%
Carbon dioxide (CO_2)	.04%
All other trace elements and gases	.06

Not included in above dry atmosphere:

Water vapor (H_2O)	~0.25% over full atmosphere, locally and large low pressure fronts 0.001%–5%

Joseph E. Schramek

Attachment II

Sulfuric and Sulfurous Acids Produced from Sulfur Dioxide-SO_2

Sulfur dioxide (also **sulphur dioxide**) is the chemical compound with the formula SO_2. It is a toxic gas and smells rotten. Naturally, it is produced by volcanic activity. In the atmosphere, it exists in trace amounts of .000002%. It is produced by burning coal and from some industrial processes. Since coal and petroleum often contain sulfur compounds, their burning can generate some sulfur dioxide. Further oxidation of SO_2 by oxygen in the air can produce the sulfurous molecule but not sulfurous acid. Aqueous solutions of sulfur dioxide, which sometimes are referred to as sulfurous acid, are used as reducing agents and as disinfectants, as are solutions of bisulfite and sulfite salts. They are also mild bleaches, and are used for materials which may be damaged by chlorine-containing bleaches. When the sulfurous molecule is wetted it reverts to SO_2 plus water that would be the equivalent to a bleaching substance. The EPA erroneously reports that H_2SO_4 Sulfuric Acid can be produced in the air from SO_3, but it requires the presence of the catalyst, NO_2 Nitrogen Dioxide. This has no chance or the remotest of possibilities of occurring in the atmosphere as both SO_3, and NO_2 gases are present at only .00005% of the air.

As a freshman at Western Michigan University, I was involving in a pre-engineering curriculum which included a 5 hour Chemistry class. Our teacher, Dr. Osborn, arranged for the class to visit a Sulfuric Acid plant in Kalamazoo. This plant is crucial to the many paper factories in place in Kalamazoo in the late 1950's. It made Aluem H_2SO_7 using sulfuric acid it produced there. It is a dehydrator and whitening agent for the manufacture of paper. I learned a lot from Dr. Osborn, and when I took my daughter, Amy, for a pre-college visitation at Western in 1994, I was surprised to see the new Chemistry Building there was named after Dr. Osborn.

In the Chemistry class there in 1957, I learned that H_2SO_4 Sulfuric Acid is very complicated to produce and I'm sharing what I learned on this process for readers of the book.

It is produced by using the **contact process** to produce the high amounts required for industrial processes. It requires the use of vanadium(V) oxide (V_2O_5) as a catalyst. The process was developed and patented in 1831 by British vinegar merchant Peregrine Phillips. In addition, the process also produces sulfur trioxide and oleum.

The process has five stages:

1. Joining sulfur and oxygen
2. Using a purification unit to purify sulfur dioxide
3. Adding oxygen to sulfur dioxide in presence of catalyst vanadium pentoxide, with temperatures of 450 degrees Celsius and pressure of 1-2 atm
4. sulfur trioxide formed is added to sulfuric acid which gives rise to oleum

5. water is added to the oleum to make very concentrated sulfuric acid.

Joining Sulfur and Oxygen

It is important to avoid catalyst poisoning and a purification unit is used to purify air and SO_2 by washing with water and drying with sulfuric acid.

To save energy, the mixture is heated by gases from the catalytic converter by heat exchangers.
Sulfur dioxide and oxygen reaction is as follows:

$$2\ SO_2(g) + O_2(g) \rightleftharpoons 2\ SO_3(g) : \Delta H = -197\ kJ\ mol^{-1}$$

According to the Le Chatelier's process, a colder temperature should be used to shift the chemical process faster to increase production. Too low of a temperature causes the process to be at a poor production levels. To quicken the production level, high temperatures (450 °C), medium pressures (1-2 atm), and vanadium(V) oxide (V_2O_5) are used to ensure a 90% plus yield. The process for the function of the catalyst takes two steps:

1. Transforming of SO_2 into SO_3 by V^{5+}:

$$2\ SO_2 + 4V^{5+} + 2\ O^{2-} \rightarrow 2\ SO_3 + 4V^{4+}$$

2. Reversal of V^{4+} back into V^{5+} by oxygen (catalyst regeneration):

$$4\ V^{4+} + O_2 \rightarrow 4\ V^{5+} + 2\ O^{2-}$$

Hot SO_3 passes through the heat exchanger and into concentrated H_2SO_4 in the absorption tower to form oleum:

$$H_2SO_4(liquid) + SO_3(gas) \rightarrow H_2S_2O_7(liquid)$$

Note that directly dissolving SO_3 in water is impractical due to the highly exothermic nature of the reaction. Acidic vapor or mists are formed instead of a liquid. Oleum is reacted with water to form concentrated H_2SO_4. The average percentage yield of this reaction is around 30%.

$$H_2S_2O_7(liquid) + H_2O(liquid) \rightarrow 2\ H_2SO_4(liquid)$$

Joseph E. Schramek

Attachment III

Sources of Nitrogen Dioxide NO_2 and Nitric Acid Coal Fired Electrical Plants

Electricity generation is a source of air emissions in the United States today. Fossil fuel-fired power plants are thought to be responsible for 67 percent of the nation's sulfur dioxide emissions, 23 percent of nitrogen oxide emissions, and 40 percent of man-made carbon dioxide emissions. These figures are impossible to determine since sulfur dioxide, nitrogen dioxide, and nitrous oxide are trace gaseous compounds in the atmosphere at levels of .00002% for each. With these concentrations in the atmosphere, the EPA's position that these emissions can lead to smog, acid rain, haze, and climate change seem at best to be doubtful or non-existent. Even with this doubt, Congress is currently considering proposals to require further reductions of emissions from power plants. Renewable energy is receiving increased attention by environmental policymakers because renewable energy technologies have significantly lower emissions than traditional power generation technologies or it may support an expensive movement to remove current power generation, equipment, and vehicles that may have not been the at fault or misjudged by the EPA.

How Nitric Acid is produced from Nitrogen Dioxide in the atmosphere

Nitric acid is made in the atmosphere by reaction of nitrogen dioxide (NO_2) with water.

$$3\ NO_2 + H_2O \rightarrow 2\ HNO_3 + NO$$

Normally, the nitric oxide produced by the reaction is reoxidized by the oxygen in air to produce additional nitrogen dioxide. Because Nitrogen Dioxide is so remote in the atmosphere at .00002%, the concentration level of the HNO_3 Nitric Acid would not be a threat to any of the population at large.

What are the effects of NO_2 and HNO_3 Nitric Acid?

NO_2 can irritate the lungs and lower resistance to respiratory infection. Sensitivity increases for people with asthma and bronchitis. NO_2 chemically transforms into nitric acid and, when deposited, contributes to lake acidification. NO_2, when chemically transformed to nitric acid, can corrode metals, fade fabrics and degrade rubber. It can damage trees and crops, resulting in substantial losses. It is doubtful that none of these events or situations could occur with the general population, vegetation, and or materials at large, since the remoteness of the created acid molecule in the atmosphere at .00002% is so insignificant in acidity. If the exposure to HNO_3 occurs in a laboratory or industrial situation where this acid was used, the severity of the exposure or contact with the acid would be greatly increase because of the higher concentrations levels of the acid in use. Likewise the HNO_3 Nitric Acid is a very aggressive oxidizer and any fabric, metal, vegetation or animal tissue in contact with the acid would be damaged and neutralize the acid causing no further damage. Therefore, HNO_3 Nitric Acid would never have a play in the ph levels of any lake, pond, stream, ocean or body of water too.

What is Nitrogen Dioxide NO_2?

Concentrated NO_2 is a reddish-brown gas with a pungent and irritating odor. This would only occur where it is stored in bulk or used in laboratory of industrial locales. If concentration levels of NO_2 were released, it would transform the air to form gaseous nitric acid and toxic organic nitrates. At low concentrations of NO_2 at .00002% in the atmosphere, it is doubtful that it would have any effects in atmospheric reactions that produce ground-level ozone, or smog, or a precursor to nitrates, which contribute to increased respirable particle levels in the atmosphere.

Attachment IV

Effects of Acid Rain - Forests

For years, some concern exists for the vegetation, and forest trees from Maine to Georgia in the Shenandoah and Great Smoky Mountain National Parks at the higher elevations over staying healthy thru the growing season. Some believe that the cause of this demise may be caused by acidic gases from coal-fired power plants and environmental stressors such as: insects, disease, drought, or very cold weather. Work to understand the issue pertains to these three areas:

Acid Rain on the Forest Floor

It is known that rain water has a ph of 5.5 which is a weak acid. Some may argue whether the acid is predominately H_2CO_3 Carbonic Acid or others. Carbonic acid is a buffer that tends to keep the acidity from increasing. The ability of forest soils to resist, or buffer, acidity depends on the thickness and composition of the soil, as well as the type of bedrock beneath the forest floor. Midwestern states like Nebraska and Indiana have soils that are well buffered. Places in the mountainous northeast, like New York's Adirondack and Catskill Mountains, have thin soils with low buffering capacity.

How Acid Rain Harms Trees?

Studies indicate that acid rain does not kill trees. Efforts seem to be centered on finding some relationship to coal-fired electric power plant emissions as a chief cause for the conditions. Others want to blame runoff of rain water for washing away key nutrients of the soil and exposure of toxic aluminum as a result of acid rain. Actually, the runoff of rain water will remove soil and nutrients necessary for growth and health of the tree. Overlooked is the need of rain water in the hot summer as a condition causing the continuation of the photosynthesis process where water and carbon dioxide provide the key elements for vegetational growth in the trees and other plant life affected. Personally, at my home, I have to ensure the trees, bushes, and other vegetation receive water and nutrients during the hot and dry days of late July and August to keep them heathy and growing.

How Acid Rain Affects Other Plants?

Some others feel acid rain may affect the health of plants. Some farmers have found that fertilizers add nutrients to the soil, and adding crushed limestone, an alkaline material, to the soil buffers the soil against increased acidity to ensure healthy growth of their plants.

Convenient Lies-The Measurement of Carbon Dioxide at Volcano Sights

NO NEED TO DO ANYTHING TO REDUCE CARBON DIOXIDE

Al Gore's book, "An Inconvenient Truth-THE PLANETARY EMERGENCY OF GLOBAL WARMING AND WHAT WE CAN DO ABOUT IT", uses carbon dioxide measurements taken at the NOAA-National Oceanic and Atmospheric Administration observatory located on the north side of the Mauna Loa Mountain at the Southern End of the Big Island of Hawaii at 4 miles and 2,600 feet below the summit of the mountain, 13,675 feet above sea level.

Since 1956 Mauna Loa Observatory has been monitoring and collecting data relating to atmospheric change, and is known especially for continuous monitoring of atmospheric carbon dioxide (CO_2). According to the NOAA, Mauna Loa is the world's oldest continuous CO_2 monitoring station, and the world's primary benchmark site for measurement of the gas. Mauna Loa was originally chosen as a monitoring site because, located far from any continent, the air was sampled and is a good average for the Central Pacific, and being high above the inversion layer where most of the local effects are present and there was already a rough road to the summit built by the military.

The CO_2 contamination from local volcanic sources is sometimes detected at the observatory. Technicians at the observatory identify these contamination events and the effects are removed from the background data. The observatory records show an increase in the CO_2 concentration from 330 ppm (parts per million) in 1975 to 400 ppm in 2014. They continued their efforts in spite of the fact that Mauna Loa is an active volcano which erupted in 1950, 1975, and 1984, and with between minor eruptions and venting of volcanic gases continuously and the continuous and ongoing volcanic activity a few miles to the Southeast on the Hawaii Island where carbon dioxide has been continuously emitted into the local atmosphere.

This lie, that measuring ambient CO_2 concentrations along with volcanic gaseous releases below the mountain top of an active volcano is OK, is used by Al Gore to show the key point that the level of carbon dioxide is increasing across the world, and alleges the change causes perception of climate change or global warming.

Fallacy of the Green Movement and Climate Change

CO_2 Concentration Measurements at Mauna Loa Mountain And Other NOAA Observatory Sites in the Pacific

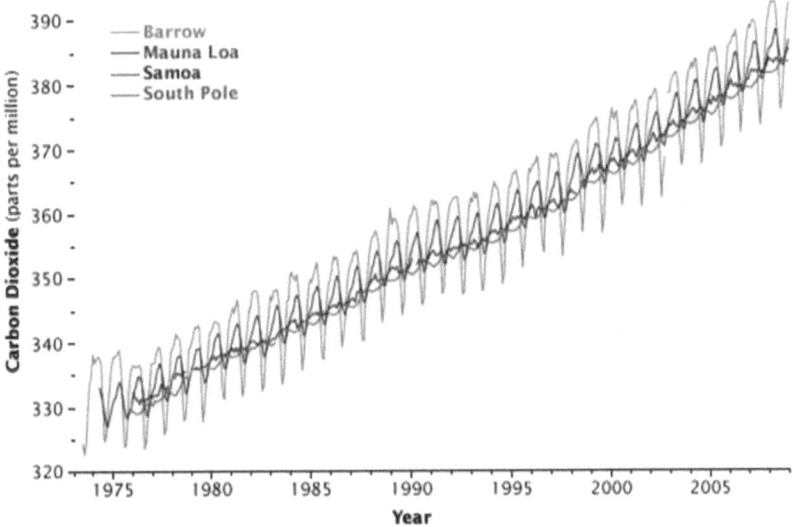

The chart above shows the growth in CO_2 Carbon Dioxide at Mauna Loa and the other NOAA observatories in the Pacific Ocean.

Some Reasons to Support the Premise That CO_2 Concentration Measurements at Mauna Loa Mountain Are In Error

- Magma contains gases that are released into the atmosphere during eruptions. Gases are also released from magma that either remains below ground (for example, as an intrusion) or is rising toward the surface. In such cases, gases may escape continuously into the atmosphere from the soil, volcanic vents, fumaroles, and hydro-thermal systems. The most abundant gas typically released into the atmosphere from volcanic systems is water vapor (H_2O), followed by carbon dioxide CO_2, and sulfur dioxide (SO_2). Other gases are released in smaller amounts. Since the carbon dioxide CO_2 curve moves upward over time, did the scientists mislead by assuming they were correct in removing those results attributable to major changes or did they ignore the smaller increases due to the other lesser volcanoes releases that cumulatively added up to the whole increase over time.

- In 1958, Charles David Keeling with the support of Harry Wexler of the U.S. Weather Service and Roger Revelle of the Scripps Institution of Oceanography, began his monumental task to measure the accumulation of carbon dioxide in the atmosphere as a debate among scientists existed whether the burning of fossil fuels would end up in the atmosphere or be absorbed by the ocean. Swedish scientist, Svante Arrhenius suggested that human-produced carbon dioxide would build up in the atmosphere and he went further to hypothesis that an accumulation of carbon dioxide would gradually raise earth's temperature. At first, Keeling found that the concentration of carbon dioxide peaked in May and then started to decline in October. He realized the rise and fall of carbon dioxide levels might be due to the regular seasonal cycle of growth and decay of plants. Sure enough, the measurements repeated the following year. The next year he made a big discovery that the levels of carbon dioxide were higher still over the prior year. Then year after year, the levels of carbon dioxide continued to rise overall.

- Keeling never considered that the carbon dioxide increases were the result of his observatory being located near the summit of the Mauna Loa volcano. With the ongoing volcanic actions from 1984 onward to current dates and the associated carbon dioxide releases from the vents atop Mauna Loa and the additional fumes from the active Kilauea volcano just a few miles to the East of Mauna Loa volcano may have accumulated each year for the increases found over time. Since the Mauna Loa has no vegetation from the summit and past the observatory, conversion of carbon dioxide CO_2 into vegetation and oxygen from photosynthesis does not occur there and the gas being the heaviest of the principal atmospheric gases, the carbon dioxide CO_2 would collect at low spots along the mountain surface downward and accumulate over time. This would interfere with the natural flow of the oceanic atmosphere and taint any measurement taken.
- The fact that Scientists use the Mauna Loa Observatory for the determination of the concentration of carbon dioxide CO_2 has to be a major error. To alleviate any error the observatory should have been closed and a new one erected in a non-volcanic location and not even remotely close to one.
- The use of the American Samoa, a chain of seven islands to confirm Mauna Loa results is in error too as the observatory lies at the east end of the island chain of four volcanoes.
- The South Pole and Barrow, Alaska have insufficient reports over time to determine correlation with the Mauna Loa results.

- It is agreed the current range of carbon dioxide CO_2 concentration is about 392 ppm with an annual range of +3 ppm in winter months, and -12 ppm at the end of the growing seasons in the Northern and Southern Hemispheres of the world. What is not agreed is that the increase on the Big Island of Hawaii is due solely to increases of carbon dioxide concentrations from non-volcano sources worldwide outside the Big Island of Hawaii.

Al Gore's **2nd Convenient Lie** in his book, "An Inconvenient Truth-THE PLANETARY EMERGENCY OF GLOBAL WARMING AND WHAT WE CAN DO ABOUT IT", **is the omission of water vapor (H_2O) as the more prevalent greenhouse gas over carbon dioxide CO_2.**

Here is the rationale to support that H_2O Water or Water Vapor (Gas) is the only factor affecting climatic conditions on earth.

- *Why burning of fossil fuels has no effect on the atmospheric temperatures and any of the climatic related disasters?*

Burning of coal and gasoline produces water vapor and carbon dioxide. Both products are greenhouse gases. Water vapor is the more potent of the two greenhouse gases. Water vapor can reach a high of 4% of the atmosphere over the ocean, while carbon dioxide is only 392 ppm of the atmosphere with an annual range of +3 ppm in winter months and -12 ppm at the end of the growing season. The water vapor molecule has a lighter molecular weight (18) but higher amounts water vapor in the atmosphere (up to 40,000 ppm (parts per million)) for water vapor versus a relatively constant 392 ppm for carbon dioxide-molecular weight (44), and has a much larger impact (100 times) on the earth's atmospheric pressure. Since the amount of carbon dioxide is almost constant at 392 ppm, and the water vapor molecule varies from 0-4% of the atmosphere, water vapor molecule is totally responsible for our violent hurricanes and tornadoes and storm damages from high straight line winds in storms.

- The atmosphere contains the greenhouse gases in varying amounts from: water vapor at a maximum level of 4% at sea level, and carbon dioxide at a relatively constant level of 392 ppm with an annual range of +3 ppm in winter months and 12 ppm at the end of the growing season depending on the seasons in the Northern and Southern Hemispheres of the world. Water vapor or humidity can lower atmospheric pressures from 30 mm for low humidity and fair weather and 26 mm for high humidity for tornado and hurricane weather conditions. Atmospheric conditions are largely dependent on High and Low Pressure Systems. High Pressure Systems at 30 mm have clear blue skies and low humidity while Low Pressure System at 28-26 mm have high humidity and temperature conditions and gray threatening skies. Because Carbon Dioxide stays at insignificant levels in the atmosphere, 400 ppm per NOAA Mauna Loa

Observatory results, it does not have any effect on weather conditions despite the EPA ruling it a pollutant.

- ***Why is the water vapor molecule responsible for all climatic disasters?***

In high pressure systems (Barometric reading of 29-30 inches of Mercury), we experience good weather with few clouds and fresh air with the highest percentage of oxygen-21% and lowest humidity in the air for comfortable breathing. In low atmospheric pressure systems (Barometric reading of 26-28 inches of Mercury) the air is very warm and humid-3% water vapor and with reduced oxygen-about 20% in the atmosphere for poorer breathing conditions and discomfort as perspiration condenses on skin without evaporation and cooling of the skin on our body. Low pressure systems are sometimes incorrectly labeled heavy air. The low pressure systems provide the water molecules which produce rain when the system is confronted with a colder temperature high pressure front that generally comes from the Northwest in the Northern Hemisphere. These fronts produce the most devastating tornadoes and thunderstorms with high straight line winds. For some reason, the scientists ignore the water vapor molecule as the cause for all the climatic temperature disasters and global warming (real or unreal or insignificant). Because there is increasing burning of fossil fuels in the world, they incorrectly choose the other greenhouse gas, carbon dioxide as the culprit, even though the percentage of carbon dioxide in the atmosphere is constantly in the range of 392 ppm.

Fallacy of the Green Movement and Climate Change

- ***What keeps the percentage of carbon dioxide relatively constant at 392 ppm?***

What keeps the percentage of carbon dioxide relatively constant at 392 ppm is that plant life has a voracious appetite for the carbon dioxide and water molecule to create vegetation to cover the earth's soil and provide food supplies for animal and human life. A simple analysis is that the voracious appetite of plant life for carbon dioxide is matched almost equally to the creation of carbon dioxide through burning hydrocarbons for heating and energy for manufacturing processes.

- ***Why is the burning of fossil fuels and hydrocarbons important for continuation of life here on earth?***

Hydrocarbons in nature include all plant and animal life, and fossil fuels (oil, and coal). Hydrocarbons are organic compounds that are produced naturally from water and carbon dioxide through photosynthesis. Over eons of years, the trapping and decaying of these remnants through seismic activity, transformation and relocation of the earth's continental plates and oceans results in the finding them deep in the earth as fossil fuels; Including: coal, and oil. Combustion of coal and oil products (gasoline, diesel fuel, and other petroleum variations) for heating and energy sources by burning, and digestion of food products by man and animal life results in re-creation of the water vapor molecule and carbon dioxide molecule. As an example, an automobile burning one gallon of gas will produce 1 gallon of liquid water when the water vapor is condensed into water. It also produces copious amounts of carbon dioxide molecules that produces and maintains our Oxygen supply for human and animal life. Without Oxygen, we could not ignite fossil fuels and waste to produce energy for survival here on earth. Any efforts to reduce the percentage of carbon dioxide in the atmosphere could eventually reduce the earth's available vegetation for animal food supplies.

- *Why is it important to maintain the available Oxygen levels in our atmosphere at 21%?*

The most devastating effect of the Green Movement would be the loss or reduction of the Oxygen molecule levels in the atmosphere from a rather constant level of 21% to lower levels. High humidity reduces the Oxygen molecule count in the atmosphere as the water molecule causes the atmosphere pressure dropping from 30 inches of Mercury to 26-27 inches for near hurricane and tornado weather conditions. Some have difficulty with breathing under these conditions especially for seniors and any workers doing heavy labor. High humidity prevents workers from evaporating their body sweat to cool their body and blood and in some cases heat strokes may occur. Oxygen molecules are only added to the atmosphere from plant life and vegetation during the process of photosynthesis. If the current Green and Carbon Footprint Movements ever reach its desired goals for significantly reducing the carbon dioxide molecule percentage of .0392% or 392 ppm levels, humans and all animal life would begin to experience breathing issues causing many variations of health concerns including possible death. The Oxygen molecule in nature only comes from the Carbon Dioxide molecule through the photosynthesis process that produces our vegetation and plant life on earth.

Summary

The evidence presented here confirms that carbon dioxide is not a pollutant but necessary to continue life here on earth and it does not have any effect on climatic events as purported by Al Gore's book, "An Inconvenient Truth-THE PLANETARY EMERGENCY OF GLOBAL WARMING AND WHAT WE CAN DO ABOUT IT". Carbon dioxide is the most important gas in our atmosphere and environment, because it gives us the only natural source of oxygen, vegetation, and plant life for food through the process of photosynthesis to enable human and animal life here on earth to continue.

NO NEED TO DO ANYTHING TO REDUCE CARBON DIOXIDE!!

Letter to President Lou Anna K. Simon of Michigan State University, "Methane Gas Versus Coal-Powered Power Plants",

Dear President Lou Anna K. Simon

Subject: Decision to Terminate Coal Burning at MSU Power Plants

On a recent WJR radio program related to making Michigan Greener, you reported that MSU would be completing its changeover from coal burning to using methane or natural gas CH_4 to reduce carbon dioxide emissions. I found this interesting as I'm about to publish a book, "Fallacy of the Green Movement and Climate Change. To show my point, I'll show the elemental picture of the products of the two types of combustion. Keep in mind that principal greenhouse gases are Water Vapor-H_2O and Carbon Dioxide-Co_2. My book shows that all major climatic disasters are related to the Water Vapor molecule that can reach 4-5% in the atmosphere in major low pressure fronts, and the Carbon Dioxide molecules are shown mostly in the range of 392-400 ppm or .04% of the atmosphere. Carbon Dioxide is the magical gas that through the photosynthesis process gives us our only natural source of oxygen and vegetation and energy products for continued life here on earth for all humans and animal life.

The Carbon content for coal ranges from 91.5% for Anthracite Coal to 50% Carbon content for Lignite Coal. Therefore, Anthracite Coal would produce the most Carbon Dioxide-CO_2 and the least amount of Water Vapor-H_2O. Where Lignite coal would produce the least amount of Carbon Dioxide-CO_2 and the most amount of Water Vapor-H_2O. I'm not aware of the types of coal you are using at your MSU power plants.

Let's look at this way:

Anthracite Coal

$C + O_2 = CO_2$ (Plus a molecular portion of the water in raw coal-1-3%)

Lignite or Brown Coal

$C + O_2 = CO_2$ (Plus a molecular portion of the water in raw coal-15-30%)

Methane or Natural Gas

$CH_4 + 2O_2 = CO_2 + 2H_2O$ (Assumes water in CH_4 removed)

Joseph E. Schramek

Overall most experts say Methane or Natural gas reduces greenhouse gases by 45% over coal burning, but this doesn't include the creation of 2 molecules of water vapor created from the combustion of one molecule of CH_4. Including the 2 molecules of water created, the methane combustion produces 50% more greenhouse gases than both types of coal burning.

Both types of coal and methane gas produce the same amount energy at the molecular level, so there is no advantage for either process. The proponents for using methane gas hide the fact that it creates more greenhouse gases than coal.

My book shows that both types of greenhouse gases, water and carbon dioxide, are essential for continuing life here on earth for humans and animals.

Respectively,

Joseph E. Schramek
MSU-1961 Electrical Engineering Graduate

Responses to Various Claims of the Environmental Movement

Man-Made Increases of Carbon Dioxide CO_2 Causing Increasing Acidity of Oceans

Environmentalist's feel that rain water mixes with Carbon Dioxide to form carbonic acid. Keep in mind that the ratio of Carbonic Acid to Carbon Dioxide is .0017 in pure water and .0012 in seawater. Hence the majority of carbon dioxide is not converted into Carbonic Acid when Carbon Dioxide is dissolved in water. The current pH level of oceans is 8.104 is a base and not acidic. Since most of the Carbon Dioxide mixed with water in rain is not acid but bubbles of Carbon Dioxide. Like a buffer, it makes sense that the Carbon Dioxide bubbles would rise out of the water due to turbulence and re-enter the atmosphere again and even dilute most of the Carbonic Acid until more rain water falls in the ocean.

It seems to me that the Environmental Scientists who predict doom and gloom for the ocean and it's marine organisms from projected increases of acidic hydrogen ions from carbonic acid in the oceans as the rise of Carbon Dioxide gas in the atmosphere increases to a pH level of sea water of 7.949 for 2050 and 7.824 in 2100. Keep in mind these values are non-acidic but base or alkaline water.

Carbon Dioxide Measurements at the NOAA-National Oceanic and Atmospheric Administration Recording Site at the 11,000 feet Facility on Mauna Loa Volcano Site

The level of 405 ppm or about 2 ppm per year increase for Carbon Dioxide levels in the atmosphere as measured at the NOAA Facility at the 11,000 feet level of Mauna Loa is backed up by other stations reporting similar results. I have doubts about the validity of Mauna Loa's results when both Mauna Loa Volcano location and the currently very active Kilauea Volcano about 18 miles East of their station are spewing carbon dioxide into the atmosphere and along the lava flow to the Ocean. I understand the high and low points for the summer and winter season in both hemispheres, but this could have been intentionally done to get acceptance of the data by other Scientists. It bothers me that all the data rises in the same proportionality for each year and especially the Mauna Loa data. I read how Mauna Loa adds and subtracts reported data as they only use a time period in the day where the Sun reaches a maximum level of the day so the atmosphere at sea level is heated such that it climbs up the slope to their station at 11,000 feet above ground level on the island. Some days, they can't use the data because the values are too low. What's worse, the station has a hard lava surface surrounding their station and down the mountain a few miles or so. There is no vegetation locally there. They claim they remove the Carbon Dioxide emitted from the vents and fissures of the Mauna Loa and Kilauea volcanoes. In trying to find other sites measuring Carbon Dioxide, it always comes up using the Mauna Loa site's values. I've learned that you should never measure the magnitude of a condition when a condition appears close by emitting the same gas. Scientist's feel it can be subtracted from the real data. I've included a couple of sheets from the NOAA Station's procedures on Pages 5-6 in the Appendix of this book, which they use to correct the data for publication. I vote for closing the facility because of the active Kilauea volcano spewing sulfur and carbon dioxides contaminating their results.

Environmentalists Claim a Warmer Climate at the South Pole Causes More Ice to Form

Environmentalists claim that the ice at the South Pole is increasing in warmer weather because the ice (pure water) dilutes the ocean salt water and the ice freezes at a higher temperature. Any ice that floats off land will never increase the water level of the ocean if melted as the ice is .9 the density of water where 90% of the ice berg is in water and 10% is above water. Any ice melting of ice on land that flows into the ocean is just like the rivers of the world that flow into the ocean and any water added to the ocean will increase the circumference of the ocean but daily evaporation of the ocean's water into the atmosphere has equalized the levels of the oceans forever. The earth can't get bigger because it is stuck with the same number of atoms over time with the exception of Hydrogen gas. Few Hydrogen gas molecules are the only ones that can escape the earth's gravitational pull. Something I learned in college. The Hydrogen molecule probably goes back to the Sun. Because of the intense gravitational pull of the earth, I believe we are stuck with the same weight, roughly the same number atoms and the current size of the earth.

Since these matters may take years to substantiate the issue or to show the issue has no merit to the degree of its sponsor's are predicting, I feel comfortable with my position. Carbon Dioxide is the miracle gas that that enables human, animal, and marine organisms to survive on earth forever until the sun is extinguished. It provides us oxygen, food, vegetation, and energy products via photosynthesis and energy from our Sun.

CO_2 Data Acceptance, Sudden Melting of Greenland's 3,000 Feet High Ice Block, Ocean Rise and Acidification Issues

A couple of years ago, I took my daughter and her children to the visit the Cranbrook Institute of Science in Bloomfield Hills. Just as we were getting ready to leave, I noticed an exhibit that declared that Michigan used to be located at the South Pole. It may be possible that the Continent of Africa was attached to South America. The air bubbles that couldn't be seen and being declared as carbon dioxide and how about how much carbonic acid was in the core samples. I thought carbonic acid is the main reason for rise in acidity of the oceans. I learned that if the molecules of gas were the same size as tennis balls, the molecules would be 15 feet apart. Then we accept all the variations of substitution and non-acceptance of some results that NOAA at the Mauna Loa Recording Station uses to make extremely exact readings of the carbon dioxide concentrations that is accepted by the World-Wide Science community. Then, we have to accept those other Measurement stations that produce readings of carbon dioxide concentration in line with Mauna Loa's findings. In summary, the only significance of the recent recording of carbon dioxide concentration pertains to the current value near .04% of the atmosphere, while the greenhouse gas water vapor climbs to 4-5% of the atmosphere and it is the only source affecting violent weather events.

The sudden melting of Greenland's 3,000 foot Ice Block on top of the island contains 2.8 million cubic kilometers of ice is only 2.5 million cubic kilometers of water as ice is .9 as dense as water. Ice never melts immediately so any melted ice is just like rivers and rain adding to the ocean. The land of the earth is only 29% of the earth's surface, and water is 71% of the earth's surface. Greenland comprises only .44% of the earth surface, and its ice volume is 2,850,000,000 cubic meters or 2,565,000,000 cubic meters of water. Sudden melting of all the ice in Greenland can't be used for calculating the rise of world's oceans as the water levels because it is gradual and simply gives the oceans more water to evaporate and create new supplies of rain water that is needed to satisfy the demands for vegetation growth, drinking water for people and animals, and refill the water needs for rivers, wells, steams, ponds, lakes, oceans and business activities. Keep in mind that the ice sheets on Greenland and the South Pole actually lowered the levels of the oceans.

It brings up a point that I included in my book in the 2nd Page of the Epilogue of my book. We learned the story of the 3 Little Pigs, that those who make homes of straw and wood, had their homes destroyed by the Big Bad Wolf who was able to blow them apart. Only the pig who made his home with bricks survived the attack and the pig lived as a result. Very few people have learned this lesson, and some of those paid their price when they were hurt or killed when a major destructive wind storm hit their home. No one wrote a book about making your home along a stream, river, rill, lake, sea or ocean where you might incur similar fate as the 2 of the Little Pigs from storm damage from floods, hurricanes, tornados, and high straight-line winds.

Summary

The earth seems to handle and react to changes of the environment from any natural of human events. I don't think there is enough respect for the voracious appetite the earth has for water and carbon dioxide to produce through the photosynthesis process: vegetation, food and energy products, and the only natural source of oxygen to enable humans and animals to continue life on earth forever until the Sun begins shut-down process.

Fallacy of the Green Movement and Climate Change

Still the hysteria continues relative to Climate Change persists:

- The US Navy expects a 15 cm rise in ocean levels for 1950 to 2000 and the Navy feels this is a top concern for them. I call this no change.
- President Obama going to a small island in the Pacific where the island had high water levels and he declares it this his top priority item to fix. Unbelievable!!
- All low-lying areas near oceans experience these events when the water is at high tides (moon and sun on the same side of earth) and the winds experiences high speed wind blows from weather fronts coming in from the ocean into the land there.
- Never build near the edges of the ocean near sea level. Go to Venice, it's waters cover the side- walks of the business districts under similar conditions.
- 10 billion metric tons of Carbon Dioxide gets dissolved in the world's ocean each year. Keep in mind, carbon dioxide doesn't affect the acidity of the ocean, the carbonic acid created is only the .12% of total of the 10 billion tons. The ocean is not acidic, it is basic with a pH of 8.069. Actually, the carbon dioxide gas is free to leave the ocean the same it does in a pop or beer. Rain water is a very weak acid of a pH of 5.65. Most will find their way into the stoma of vegetation to produce future energy, building and food products to satisfy our population of inhabitants of the earth.
- After 40 years, you can see the little stones in concrete exposed on the top surface of the side-walk in front of my house due to the acid rain.

Joseph E. Schramek

Give some credit to the earth on its natural feedback mechanisms that drives the Earth's biosphere back to equilibrium over time. Some concern existed in the Middle Ages when the population had shortages of energy and food products, which was saved by the Industrial Revolution. The book, "Fueling Freedom-Exposing the Mad War on Energy" by Stephen Moore and Kathleen Hartnett White shows how the problems with providing enough food and income for the growing population which seemed to be self-correcting with the advances in using fossil fuels and the start of the Industrial Revolution. As I mentioned before, Steve and I traded our books, but my positions on this topic were well established on my own a few years back. Steve said my book was right on. I just couldn't see why the Green Movement and Climate Change was unsupported by what I learned from with my education and my general interests in why things happen in nature. I learned a lot from pitching baseball as a youth on hot and humid and fair and clear weather days, watching airplanes land and take off and the distance for different weather conditions, and the distance a golf ball travels for various temperature and humidity conditions.

VW 590,000 2009-2016 Vehicles Equipped With 2 L and 3-L Diesel Engines with Defeat Devices-To Cheat on EPA's NO_x Emissions Requirements

What Harm Resulted or Was it False Science?

What was the result of VW using a Electronic Cheat Setting to meet Compliance to NO_2 Standards for Certification

It enabled VW for Volkswagen and Audi diesel vehicles to use their Engine Control Unit (ECU) to use a special test mode setting for government emission testing to meet the requirements of the Standards for NO_2 then switching to an alternative programming for customers driving in the real world. The EPA estimate that the cheating VW diesels polluted at up to 40 times the emissions standards for NO_2 under maximum vehicle load and throttle conditions, but under normal driving conditions the emissions were more in the range of 10 to 20 times over the federal limit for NO_2. After testing, the CPU would revert to the higher

Joseph E. Schramek

NO_2 emission levels and better engine response and fuel economy for the owner's benefit.

What is NO_x?

NO_x is Nitrogen Dioxide NO_2. NO_2 is created by the heat created from internal combustion engines burning fossil fuels. Outdoors, NO_2 can be a result of traffic from motor vehicles. Indoors, exposure arises from cigarette smoke, and butane and kerosene heaters and stoves. NO_2 is one of the rarest gases in the atmosphere at .000002%, even with the NO_2 from the above sources included. It does not pose any health threat to the human population at large.

Acute harm due to NO_2 exposure is only likely to arise in occupational settings. Direct exposure to the skin can cause irritations and burns. Only very high concentrations of the gaseous form cause immediate distress: 10-20 ppm can cause mild irritations of the nose and throat, 25-50 ppm can cause edema leading to bronchitis or pneumonia, and levels above 100 ppm can cause death due to asphyxiation from fluid in the lungs. There are often no symptoms at the time of exposure other than transient cough, fatigue or nausea, but over hours inflammation in the lungs cause edema. The US requires these occupational sites are subject to strict reporting requirements for facilities which produce, store, or use NO_2 in significant quantities.

Fallacy of the Green Movement and Climate Change

Why does the EPA include NO_2 Emissions Requirements for Automobiles and Light Trucks?

The EPA seems to be slanted to eliminate or reduce any gas emitted from any engine or power plant burning fossil fuel or the heat generated from a burning or high heat process. It started with the designation of carbon dioxide as a pollutant and the measures taken to close many Electrical Power Plants burning fossil fuels. The EPA overlooked that the carbon dioxide gas is necessary in our atmosphere to ensure the continuation of life here on earth for humans. Carbon Dioxide, along with water, and heat from the Sun provides through photosynthesis all the natural generation of oxygen for humans and animals to breath and live, the vegetation to enable plant and food products for humans and animals to eat and live, and the hydrocarbon products: lumber, oil and gases, and coal for their energy needs to survive on earth.

EPA took notice of the Northeastern States complaints of the reddish-brown hue of the marine fog or other smog type conditions and it developed a listing of the total weight of NO_2 produced by power plants, cars and trucks, and other settings where NO_2 is produced, stored, or used. Possibly the States and the EPA might have forgotten that the sky color is azure or blue from the well-known prism effect of the atmosphere that causes the sky to show different colors of the sky, clouds, smog, and other conditions of the sky. The reddish-brown is a regular sky color for early daylight when the sun rises at an acute angle in the early morning from the East and when the Sun sets at an oblique angle in the Western skies. The Western side of the State of Michigan along the Lake Michigan skyline is famous for the reddish color of the late evening sky. Because most of the population of the Northern States lives on the Eastern Coastline of the Atlantic Ocean, early morning marine fog is a regular event for them. The Western States have the same problem with early morning marine fog. These fogs clear up as the sun rises in the sky and from the increase of temperature from the sun. The reddish-brown colors of the fogs and clouds disappears as a result. Both the Northern States and the EPA took advantage of the reddish-brown hue of these smog conditions as it matches the color of the NO_2 gas. They used the charts showing the yearly States total weights of NO_2 from all sources of generation of the gas and the severe health effects of just the sites where NO_2 is produced, stored, and used and where some workers experienced inhalation of the gas by errors at these locations.

As a result, EPA began to use the health issues from those who work in occupational settings for producing, storing, and using NO_2 in significant quantities to sell their increasing standards for emission control of NO_2 for automobiles and light trucks for both gasoline and diesel equipped vehicles. EPA would always use the yearly net weight in tons of NO_2 created from all producing sources without ever mentioning that the NO_2 from all sources of creation is only .00002% of the atmosphere. It also fails to mention that NO_2 mixes with water vapor to form less than .00002% HNO_3 Nitric Acid in the atmosphere, while Carbon Dioxide CO_2 at .04% of the atmosphere also mixes with water to form less than .00005% the acidic buffer H_2CO_3 Carbonic Acid in the atmosphere. During rains, both free Carbon Dioxide CO_2 and Nitrogen Dioxide NO_2 molecules and the acids created enter the soil from nightly dew or rain to further their generation of other vegetation and food to support life here on earth.

Why did the EPA make Tier 3 Standards Beginning With 2008 Cars and Light Trucks More Stringent Than the Prior Level 2 Standards

For 2008 Car and Light Truck Models, the EPA adopted a new SFTP (Supplemental Federal Test Procedure)-US06 which was developed to address the shortcomings with the prior FTP-75 in the representation of aggressive, high speed and/or high acceleration driving behavior, rapid speed fluctuations, and driving behavior following startup.

It would appear the outcome of this change favors gasoline powered vehicles (mostly US Models) and adversely affected the performance of diesel powered vehicles (mostly European Models-Volkswagen Diesel Engine Equipped). The fact that the new SFTP-US06 Emission Test Procedure makes the Volkswagen 2L and 3L diesel quipped vehicles in violation of the EPA Standards testing, but passing when the CPU is set for passing the Emissions Standards. Apparently, VW chose to cheat on the test to protect their customer's acceptance and expectations for Best-In-Class vehicle acceleration response and fuel economy. Since the NO_2 created by the heat of combustion of gasoline and diesel fuels for cars and trucks has a short life as the NO_2

gas state is only in an interim state where this gas is quickly joined with water vapor in the atmosphere to form Nitric Acid HNO_3 molecules. For this reason, NO_2 is one of the rarest gases in the atmosphere at .000002%, even with the NO_2 from the other indoor sources included. It does not pose any health threat to the human population at large. Each molecule of Nitric Acid HNO_3 is very heavy and falls to the ground due to dew, and rain. New vegetation will develop from the ionic state NO_3 nitrates from the acid in the soil as it generates the lettuce, spinach, and other vegetables you may eat.

Time to Add Nitrogen Dioxide to the List of False Issues EPA's Has Created Over Time

Here is a list of the most prominent false science issues created by the EPA where they set Emission Standards:

- Acid rain kills trees along the Allegany Mountain Range. Acid rain is predominately Carbonic Acid from the mix of Carbon Dioxide CO_2 and water vapor during rain fall. Rain water has a pH of 5.5, a weak acid. The Carbonic Acid acts as a buffer solution maintaining the pH at 5.5 overtime.
- Carbon Dioxide is a pollutant. EPA incorrectly closed many Coal Powered Electrical Generating Power Plants. It generates our natural source of oxygen, plants, trees, food products, and our fuel and energy sources to continue life here on earth for humans and animals.
- New Natural or Methane Gas CH_4 Gas Powered Electrical Power Plants have less fossil fuel emissions

than Coal Powered Electrical Generating Power Plants. The water vapor created from burning CH_4 is a fossil fuel emission. Water vapor was the total cause of Hurricane Harvey in Houston, Texas and the flooding that occurred everywhere in Houston, and Beaumont, Texas. Therefore, the EPA is false to use Natural or Methane Gas CH_4 Gas Powered Electrical Power Plants, since they produce more fossil fuel emissions over Coal Powered Electrical Generating Power Plants.

What damage has the EPA done to the World Population and USA and Germany Manufacturing?

Both the EPA and the UBA (Germany's main environmental protection agency) of Germany have adopted similar slants on Diesel emissions. In Germany, many of their people in large Metro cities complained of the Diesel exhaust smells and odors, especially in colder weather. Without identifying the offending gas, they have adopted the position of the EPA that harm from this odor is the same as the overexposure of the NO_2 at facilities that produce Nitric Acid, store it, or use it operations at many facilities across Germany

The UBA had set emission standards for diesel equipped vehicles for Volkswagen and other companies built and operated in Germany and Europe, and none of their vehicles could meet the strict Standards either and the UBA took no action against them.

In the future, American Companies will have trouble selling their products in Germany and Europe, for the damages

Joseph E. Schramek

and financial losses to VW Volkswagen AG and worker's families form EPA's fines and penalties. VW has agreed to a $1.4 Billion Penalty to the EPA, and a $14.7 Billion Settlement overall to complete their buybacks, updates to existing vehicles, and scrappage of some vehicles programs to satisfy the EPA and California CARB.

EPA uses False Science to Condemn NO_2 Nitrogen Gas (An Interim State) in the Atmosphere Before Forming HNO_3 Nitric Acid

The Earth Uses the Nitrate Ion NO_3 from HNO_3 Nitric Acid from rain water for Fertilizer for Vegetation and Vegetables to Eat

"Water-A Deeper Understanding of Our Most Vital Resource" Article in the Fall 2017 MSU Alumni Magazine

Parking Lot View of Cloud Formation Near Pontiac

The MSU Alumni Magazine is issued Quarterly and the 2017 Winter copy focused on MSU going deeper into "Water-A Deeper Understanding of Our Most Vital Resource". I sent a reply to the Editor of the Magazine and she forwarded to the Michigan State University President, Lou Anna K. Simon, as I challenged some of the articles in the magazine. The cover page showed a picture of a sailboat in a body of water. It didn't make sense to me that this was indicated as representative of the reason water is a source of vital importance. I know a little about where that water comes from. They never mentioned that all water comes from clouds, and the water vapor created from burning hydrocarbons in transportation vehicles and trucks, human and animal respiration, decay of vegetation, and the burning of fossil fuels for heating homes and businesses. The article emphasizes that the Great Lakes States and Canada water shed is the largest on earth. I felt the article was a little shallow as it focused on the bottom part of the cover page, liquid water, and it failed to mention the fact that all the water in the Great Lakes and everywhere in the world comes from the clouds shown on the top side of the cover page. I have deep interest in the cloud formations as shown in the photo above. This photo was taken on one of my daily return trips home to Dearborn Heights from Pepino's Restaurant on Orchard Lake Road about 1 mile West of Telegraph in Sylvan Lake, Michigan. I've taken this route for the last 12 months as my son, Tom, is a Chef, at Pepino's. The many lakes and ponds situated in the Pontiac, Oakland County area ensure many of these beautiful cloud formations on routine basis, generally in the Mid-Afternoon time-period, the hottest temperature of the day.

Actually, all water in the Lakes and Rivers, and the Oceans comes from rain water that fills these sites directly or from feeder rivers, rias, streams, creeks, ponds, marshes, swamps, storm sewers and other water drains. The rainy conditions arise from weather conditions where Low Pressure fronts meet up with High Pressure fronts or Hurricane formations where a warm Low Pressure front (plenty of water vapor) of the atmosphere meet colder High Pressure front (absence of water vapor).

Rain water is acidic with PH of 5.5. Actually, the acidity of rain water comes from pure water vapor or drops of rain water liquid coming into contact with carbon dioxide in the atmosphere to form carbonic acid. All humans drink carbonated water in the form of soda, and beer drinks on a regular basis. The point is the acidic rain water keeps the carbon dioxide levels in the 400 ppm range or less as measured at NOAA-National Oceanic andAtmospheric Administration recording site on Mount Mauna Loa at 11,000 feet on the Hawaii Island the largest island of the Hawaii Islands. All the liquid water in the Great Lakes from the world's largest water shed that is not consumed or evaporated in route to the Atlanta Ocean will be mixed with the salt water of the ocean, and eventually will end up being evaporated into the atmosphere to form new clouds and weather fronts until they rain again repeating the cycle over and over.

Many are worried about keeping our water ways clean when in fact the Great Lakes is so clean you can drink it with no problems as long as you are away from the shoreline at distance. The worry of contaminating the fish with chemicals and elements like mercury may be overstated in that fish and humans only lets matter at the ionic state into their blood and any particulate dirt is excreted or remains as a particulate inside the fish or human. Anybody who

Joseph E. Schramek

eats fish regularly may experience higher levels even at the elemental and ionic state of the contaminate.

I'm a little concerned over the large effort MSU is planning on the study of water. All the vegetation on earth is formed from water and carbon dioxide through photosynthesis. After billions of years, I believe nature will survive with little or no human efforts to preserve it for future inhabitants.

I hope you may consider some of my points on this subject.

Flint Water Contamination from Switching from Detroit Water Supply To Water Originating in the Flint River

Background

Flint built its first water treatment plant (now defunct) in 1917. The city built a second plant in 1952.

At the time of Flint's population peak and economic height (when the city was the center of the automotive industry), Flint's plants pumped 100 million gallons (380,000 m^3) of water per day. With the decline of the city's industry and significant drop in the city's population (from almost 200,000 to about 90,000 today), Flint pumped less water. By October 2015, when the Flint plant ended full time operations again, it pumped just 16 million gallons (61,000 m^3) daily.

Historical Time Line for Flint's Water Supply

- Prior to 1967, Flint relied on the Flint River for the source of water.

Joseph E. Schramek

- In 1967, Flint switched to water treated, processed, and supplied by the DWSD-Detroit Water and Sewerage Department because of poor quality and toxic substances with the water supplied by the Flint River for many years. The new water supply from Lake Huron comes from the intake 5 miles from the shore line at Port Huron at the mouth of the Detroit River. DWSD provides high quality drinking water for 40% of the state's population, serving 126 southeast Michigan communities.
- In 2011, the state of Michigan took over Flint's finances after an audit projected a $25 million deficit.
- March 22, 2012, Genesee County announces a new pipeline to provide is being designed to provide water directly from Lake Huron to Flint. The plan is to reduce costs by switching the city's water supplier from the DWSD to the Karegnondi Water Authority (KWA).
- April 16, 2013, on the city council's recommendation, Andy Dillon, state treasurer, authorizes Flint to make the switch to use water from the Flint River. One day later, the DWSD terminates its water service contract with Flint, effective April of 2014.
- During 2014 to reduce the water fund shortfall, they switched water sources from DWSD to the Flint River, while a new pipeline connecting Flint directly to the Lake Huron water intake was under a two-year construction period transition. In January 2015, users toted jugs of discolored water to a community forum.
- June 24, 2015, An EPA manager issues a memo, "High Lead Levels in Flint," warning the city is not providing corrosion control treatment to mitigate the presence of lead in drinking water.

- October 16, 2015, Flint switches back to Detroit-DWSD water supply because of numerous issues with the quality of the Flint River water for dangerous elevated lead levels at two to three times higher than the water supplied by DWSD. DWSD water tested at the 90 Percentile value of 2.3 ppb, well under the actionable level of 10 ppb for DWSD and the EPA limit of 15 ppb.
- January 13, 2016, Governor Snyder announces outbreak of Legionnaires' disease occurred in the Flint area between June 2014 and November 2015, with 87 cases and 10 deaths. It is unclear, however, whether the spike is linked to the water switch.
- January 24, 2017, The Michigan Department of Environmental Quality says the lead levels in the city's water tested are below the federal limit in a recent six-month study. This is with the Detroit-DWSD water supply.
- Currently, the decision to determine the future source of Flint's water supply rests in the Detroit Federal Court-Judge David Lawson, on whether to continue use of the Detroit-DWSD water via a lease with the Great Lakes Water Authority or switch the water supply from Lake Huron via the Karegnondi Water Authority which requires tens of millions of dollars for repairs and updates the Flint treatment plant and 3 ½ years for the update.

Joseph E. Schramek

Key Health Issues and Possible Diseases Resulting from Using Flint Water

Some of the problems with using Flint Water raised concerns over the following issues:

- Acidity of Flint water and corrosion of the city water mains and the lead pipes from the mains to the individual home sites where some children and some adults experience high levels of lead in their blood.
- Lead from other foods-Meats, fish, drinks, and vegetables
- Outbreak of Legionnaire's Disease causing 10 deaths and 87 cases
- Users delivered bottles of discolored water from their water taps in their home to a community forum.
- Need to continue using water from plastic bottles for drinking.

Acidity of Flint Water and Corrosion of the Water Mains, and Lead Pipes to the Homes

All rain water that enters the Flint river has a pH of 5.5, a very weak acid, carbonic acid, that is created from contact with Carbon Dioxide in the atmosphere combining with water vapor in the clouds or as the rain falls to the ground. The ratio of carbonic acid to water is only 1.7×10^{-3}. It maintains this proportion like a "buffer" solution, e.g. The surface of your side walk takes about 20-30 years to remove the top level of the cement material of the side walk to dissolve from the weak carbonic acid in rain water when the stones in the concrete begin to show. Carbonic acid is a weak acid and a buffer solution that retains its weakness because the acidic content remains at the low-level ratio 1.7×10^{-3}, over use and time.

The use of lead for water pipes is problematic in areas with soft or acidic water. Hard water forms insoluble layers in the pipes whereas soft and acidic water dissolves the lead pipes. The harder the water the more calcium bicarbonate and sulfate it will contain, and the more the inside of the pipes will be coated with a protective layer of lead carbonate or lead sulfate. Since lead were only used in the Flint systems for some castings along the main line, and the short connection from the water main to the user's home, or other users, it would be minimal contribution overall versus the issue here. Lead pipes oxidize when left in open air and it protects the inner pipe surfaces. Lead, itself, resists sulfuric and phosphoric acids, but not hydrochloric or nitric acids. Lead pipe was used in most homes from the water main to the user's home or business facilities as it was more corrosive resistant than steel pipes and it prevents flooding, if the pipes would leak or break from corrosion.

During October 2014, GM's Engine Parts Plant in Flint had to stop using Flint Water because a problem with too much chlorine in the water. It was corroding their production engine parts. They arranged to buy water from a neighboring township that got its water from Lake Huron via Detroit-DWSD water supply. Chlorine or Hydrochloric Acid probably was used by the Flint Water facility when it was getting its water supply from the Flint River. Chlorine is presently an important chemical for water purification (such as in water treatment plants).

Children and Some Adults with High Lead Levels in Their Blood

Lead has no confirmed biological role in the body. Its prevalence in the human body-at an average of 120 mg and is exceeded by only by Zinc (2,000 mg), and iron (4,000 mg) among the heavy metals. Lead salts are very efficiency adsorbed by the body. A small amount of Lead (1%) is stored in bones, the rest is excreted in urine and feces within a few weeks of exposure. Only a third of lead is excreted by a child. Continual exposure may result in the bioaccumulation of lead.

Lead is a highly poisonous metal (whether inhaled or swallowed), affecting almost every organ and system in the human body. At airborne levels of 100 mg/m3, it is immediately dangerous to life and health. Most ingested lead is absorbed into the bloodstream. The primary cause of its toxicity is its predilection for interfering with proper functioning of enzymes. It does so by finding to the sulfhydryl groups found on many enzymes, or mimicking and displacing other metals which act as cofactors in many enzymatic reactions. Among the essential metals that lead interacts with are calcium, iron, and zinc. High levels of calcium and iron tend to provide some protection from lead poisoning, low levels cause increase susceptibility. Lead poisoning is responsible for many symptoms of many illnesses and health issues in the body.

Until recently, children were identified as having a blood lead level of concern if the test result is 10 or more micrograms per deciliter of lead in blood. Chelation therapy is recommended for children is found with a test of greater than or equal to 45 micrograms per deciliter of lead in blood. Treatment for lead poisoning normally involves the administration of Dimercaprol and Chemet (Succimer). Acute cases may require the use of disodium calcium edetate, the calcium chelate of the disodium salt ethylene diamine tetra-acetic acid (EDTA). It has a greater affinity for lead than calcium, with the result that lead chelate is formed exchange and excreted in the urine, leaving harmless calcium.

The Flint River drains a water shed of 1,332 square miles (3,450 km^2) of Michigan, in Lapeer, Genesee, Shiawassee, Saginaw, Oakland, Tuscola, and Sanilac counties In the city of Flint, the river flows past the sites of several former General Motors factories, most notably Chevrolet's first assembly plant, which was bisected by the river, and downtown through the campus of the University of Michigan–Flint and Riverbank Park. Also along the river front is the Flint Carriage Factory site, later Dort Motors. The Durant-Dort Carriage Company Office, now a historic landmark, is across the street. Continuing downstream, the river runs past Kettering University and McLaren Hospital, then into Flint Township and through Flushing. The stretch of the Flint River from downtown Flint to Kettering University is channelized with concrete sides.

Joseph E. Schramek

Lead from Other Foods-Meats, Fish, Drinks, and Vegetables

All the blood tests run on children in Flint included lead levels from the foods they consumed up to the date of the test. These are shown in the Appendix Pages 166-171. These may have some effect on the child development issues and some mental issues with aging adults, yet never considered by the medical community, especially those who favor specific controlled diets for health.

Legionnaire's Disease with 87 Cases and 12 Deaths from June 2014 and November 2015

Legionnaire's disease is a severe form of pneumonia-lung inflammation usually cause by infection from a bacterium known as legionella. You can't catch the disease from person to person contact. Instead, most people get the disease from inhaling the bacteria. Older adults, smokers, and people with weakened immune systems are particularly susceptible to the disease. ***Pontiac Fever,*** which is caused by various species of Gram-negative bacteria in the genus ***Legionella,*** was first identified on several workers at the County's Department of Health of Flint in 1968. The workers came down with a fever and mild flu symptoms, but not pneumonia. After the 1976 Legionnaire's outbreak in Philadelphia, the Michigan health department re-examined blood samples and discovered the workers had been infected with the newly identified ***Legionella pneumophila.***

The bacterium Legionella pneumophila is responsible for most cases of legionnaires' disease. Outdoors, legionella bacteria survive in soil and water, but rarely cause infections. Indoors, though, legionella can multiply in all kinds of water systems-hot tubs, air conditioners, and mist sprayers in grocery store produce departments. Although it's possible to contract legionnaires' disease from home plumbing systems, most outbreaks have occurred in large buildings, perhaps because complex systems allow the bacteria to grow and spread more easily.

Most people become infected when they inhale microscopic water droplets containing legionella bacteria. This might be the spray from a shower, faucet or whirlpool, or water dispersed through the ventilation system in a large building. Two other ways the infection can be transmitted are:

- **Aspiration**-This occurs when liquids accidentally enter your lungs usually because you cough or choke while drinking. If you aspirate water containing legionella bacteria, you may develop legionnaire' disease.
- **Soil**-A few people have contracted the legionnaires' disease after working in the garden or using contaminated potting soil.

It is important to find the real reason for the outbreak of the Legionnaires' disease in Flint to prevent further reoccurrence of this deadly disease for the people of Flint. It is equally important to determine if the switching of water from Detroit DWSD to the Flint River and back to the Detroit DWSD had any bearing on this issue.

Despite fact that no direct cause for the Legionnaires' disease had been determined relating the switch over to Flint water from April 2014 to October 2016, Michigan Attorney General Bill Schuette charged Michigan Health Director Nick Lyon and 4 others with involuntary manslaughter and misconduct in office, both felonies. Mr. Lyon, 49, of Marshall is accused of causing the death of Robert Skidmore, on December 13, 2015 by failing to alert the public about a foreseeable outbreak of legionnaires' disease. It's a 15-year felony. Mr. Skidmore was 85 at the time of his death.

Users Delivered Bottles of Dirty Water from Their Water Taps to a Community Forum

During 2014, Flint announces fecal coliform bacterium has been detected in the water supply, prompting a couple of boil water advisories Flint residents. After flushing the pipes and adding more chlorine to the water, the city announced they can safely resume drink the water. Because Pontiac's water supply was switched a few times between Detroit-DWSD and the Flint River water sources, the water in the supply mains and the lead pipes from the mains to the home and other sites had sediments formed on the inside of the mains and pipes over time, when the switches were made the water in the mains and pipers reversed directions and loosened up the sediments and they appeared in water at the open taps being used in the homes of Flint. Since the whole water system was affected in Flint, the discoloration could continue for extended time and repeat each time the water source was switched between the two sources. The State of Michigan provided free water filters to Flint residents to clean water of sediments and lead.

Need to Continue using Water from Plastic bottles for Drinking

November 16, 2016, the State of Michigan and the city of Flint were ordered to provide free plastic bottled water to homes in Flint, where the government hasn't checked to ensure that filters are working properly. In court documents, the leader of a nonprofit group helping residents said that as many as 52% of the water filters installed in a sample of more than 400 homes had problems. Flint received millions and millions of free bottled drinking water from special interest groups, wealthy individuals, and Government agencies. It is estimated that 84% of the empties will be added to State City Solid Waste Landfill sites, and the remainder 16% will be recycled.

Summary

The Flint City Council on Sunday, October 22, 2017 asked a federal judge to reconsider an order forcing it to choose a long-term water source, a day before the panel was ordered to decide. In an emergency motion Sunday, the City Council told U.S. District Judge David Lawson that if the court doesn't dismiss or reinstate the case, the city would "forced under duress to decide on a long-term contract" before it has an expert analysis. The city is considering three choices: The Great Lakes Water Authority-Detroit (DWSD)-The current supplier of Flint's water supply, the Karegnondi

Joseph E. Schramek

Water Authority which requires 10's of millions in repairs and updates to the current Flint Water Plant and 3 ½ years to complete, and the third would require changes to the current Detroit-(DWSD) proposal, if required by the expert, Gary Cline.

Because Flint is currently in a phase of poor economics with its declining population and failure of some residents to pay their water bills-only 49% do, it would appear the Great Lakes Water Authority-Detroit (DWDS) is the only sensible choice for all until the future of Flint improves.

Mankind since the beginning of time dealt with problem of finding and using water for drinking. The following are some alternatives used by man and animal life effectively in years past:

- Since Adam and Eve, humans quickly learned how to make a fire, and this probably was used to boil water to ensure it was OK to drink.
- Taste, smell, and clearness are key measures of good water for drinking.
- During a wilderness trail walk at the Michigan Kent Lake Recreational Park with a group of Cub Scouts under my control as Cub Master, the tour guide mentioned that the safest water to drink if lost in the forest is to take a drink from stagnant pond, since the adjacent vegetation purifies the water for drinking.
- All wild animals drink water from ponds, streams, rivers, rias, and lakes to satisfy their thirst.

- Our farm was supplied by water from a water well just 40 feet from our farm house. Most rural properties in the United States and around the World get their water from these wells.
- The Great Lakes are the source of the largest supply of fresh water in the world. You can drink it without any issues when taken from open water away from the shoreline, the same as the Great Lakes Water Authority does for the Detroit (DWSD) Water Plant that Flint currently uses. It comes 5 miles offshore just North of the Port Huron at the mouth of the Detroit River. The water is fed by Lake Superior, Lake Michigan, and Lake Huron.

Letters to the Editor of the Dearborn Press and Guide-May 7, 2011 "Global Warming-Real or Imagined"

Support for Connolly comments on global warming, disputes letters from Colovas and Roush

Re: Letters by Dan Colovas and Matt Roush commenting on John Connolly's article on "Global Warming-Real or Imagined?

As the Earth rotates around the sun at 67,000 mph, and spins in the Detroit area about its axis at 750 mph, I suggest these two gentlemen stop reading books written by climatologists and enroll in chemistry and physic classes to get a better understanding of the elemental factors involved on this topic. I'm submitting my positions that challenge these two gentlemen and most of the great analyzers of today, the effect of carbon dioxide gas on global warming. My positions are based on common sense and elemental science.

Recent fear reports that Al Gore's projected rise of the ocean levels from melting of the polar ice caps from global warming would require levees to be built at major U.S. cities on the Atlantic and Gulf Coast lines. All the reports are untrue. Liquid water, icebergs, and humidity consist of water molecules-H_2O or two hydrogen atoms and one oxygen atom for a total molecular weight of 18.

The amount of water in its many forms is a fixed amount on the planet Earth. Most fear mongers mention the melting of polar ice cap adds to the ocean sea level. The fundamental error is that an iceberg is .9 the density of water, and therefore it floats with .9 of its volume below the ocean level and .1 of its volume above sea level. When an iceberg melts its density is equal to water. Therefore, when the iceberg melts it merely replaces the submerge level of the iceberg and since its density in the melted form is equal to adjacent ocean sea level, and there is no rise in the overall level of the ocean.

Additionally, most scientists overlook that carbon dioxide is the magical gas that creates our oxygen from the process of photosynthesis. Photosynthesis marries carbon dioxide, water, and other nutrients in our soil to produce vegetation, trees and oxygen. The plant-life on Earth has a veracious appetite for carbon dioxide and this keeps the level of this gas at non-threatening levels for humans and adds the necessary oxygen levels to support human life. Fortunately, carbon is the most prevalent atom in support of our life. It's the principle atom for the manufacture of sugar, wood, cloth, plastic, coal, coke, charcoal, diamonds, carbohydrates, gas, hydrocarbons, etc.

Joseph E. Schramek

Also, the principal leaders of the global warming have developed a situation where people are designated as believers in it or non-believers. If a situation is relegated to one where a person either believes or doesn't believe, it questions the accuracy of either argument. Since neither side is a majority, I conclude that the situation is: true, not-true, or not a significant effect. My basic and elementary statements above support a situation where the total Earth systems in place have the capability of handling the ranges of current variations of climate, and absolute levels of oxygen, hydrogen, carbon atoms that the Earth systems contain without any threats to our existence. Thank God the Earth spins around sun. Amazingly, the sun only uses the most insignificant amount of its radiant energy to satisfy the needs of Earth's inhabitants until the end of time.

It's so simple. Why can't scientists or climatologists comprehend it?

Letter To the Editor of the Detroit News-May 4, 2011

Subject: Nation and World-Arctic study warns of melt-Associated Press Article by Karl Ritter and Charles J. Hanley

Scientific panel predicts a 5 foot raise in sea levels by end of century from melting of Greenland, and the polar ice caps from accelerating global warming. This is impossible! The amount of water in its many forms is a fixed amount on the planet Earth. Most fear mongers mention the melting of polar ice cap adds to the ocean sea level. The fundamental error is that an iceberg is .9 the density of water, and therefore it floats with .9 of its volume below the ocean level and .1 of its volume above sea level. When an iceberg melts its density is equal to water. Therefore, when the iceberg melts it merely replaces the submerge level of the iceberg and since its density in the melted form is equal to adjacent ocean sea level, and there is no rise in the overall level of the ocean. The scientists believe the acceleration of global warming is attributable the bogged down efforts to reduce carbon dioxide emissions over the past two decades.

Joseph E. Schramek

Most scientists overlook the importance of carbon dioxide in sustaining life on earth. Carbon dioxide is the magical gas that creates our oxygen from the process of photosynthesis. Photosynthesis marries carbon dioxide, water, and other nutrients in our soil to produce vegetation, trees and oxygen. The plant-life on Earth has a veracious appetite for carbon dioxide and this keeps the level of this gas at non-threatening levels for humans and adds the necessary oxygen levels to support human life. Fortunately, carbon is the most prevalent atom in support of our life. It's the principle atom for the manufacture of sugar, wood, cloth, plastic, coal, coke, charcoal, diamonds, carbohydrates, gas, hydrocarbons, etc.

Also, the principle leaders of the global warming have developed a situation where people are designated as believers in it or non-believers. If a situation is relegated to one where a person either believes or doesn't believe, it questions the accuracy of either argument. Since neither side is a majority, I conclude that the situation is: true, not-true, or not a significant effect. My basic and elementary statements above support a situation where the total Earth systems in place have the capability of handling the ranges of current variations of climate, and absolute levels of oxygen, hydrogen, carbon atoms that the Earth systems contain without any threats to our existence.

Thank God the Earth spins around sun at 67,000 mph and rotates at 1000 mph at the equator. Amazingly, the sun only uses the most insignificant amount of its radiant energy to satisfy the needs of Earth's inhabitants until the end of time.

It's so simple. Why can't scientists or climatologists comprehend it?

MSU Gets Wet-Letter to Editor of the MSU Alumni Magazine

Response to an article, "MSU Gets Wet", in the Spring, 2012 publication of the MSU Alumni Magazine that has a distribution to 40,000 plus MSU Alumni.

 IN **BASKET**

MSU GETS WET

Your cover story seems to ignore that water usually comes from rain that comes from the condensation of water vapor that comes from the evaporation of the ocean, sea, river, lake and pond supplies of liquid water, exhalations of animal and human life, and burning of hydrocarbons. Hydrocarbons are all plant life and fossil fuels.

The statement, "Water is not the new oil, it is far more important. It sustains life," is incomplete. It should have included carbon dioxide. Water in the vapor form and carbon dioxide in the gaseous mode are the two most prevalent greenhouse gases. Carbon dioxide gives us oxygen from photo-synthesis for creating vegetation. Carbon dioxide is relatively constant at 392 parts per million or .0392% of our atmosphere. The earth's vegetation has a voracious appetite for this gas. To get more fresh water, humans need to burn more fossil fuels and vegetation to get more water vapor and carbon dioxide in the atmosphere.

Joe Schramek, '61
Dearborn Heights

Response to "earth is dying" and "Carbon Diet and Footprint" Advocates

Dear "earth is dying" and "Carbon Diet and Footprint" Advocates,

Many advocates will tell you "water is blue gold" and only 3% or less is available as so called fresh water for refinement to purer water for human's consumption. They overlook that all water, both salt and fresh, comprising the other 95% of available water is composed of (2 Parts Hydrogen and 1 part Oxygen) and it already exists in oceans, polar ice caps, seas, lakes, streams, rills, and ponds of the surface of the earth and also in deposits in the earth where we retrieve with pumping wells...etc. The evaporation of these liquids collects in the atmosphere and forms higher temperature low pressure fronts that contain huge amounts of water vapor that reduces the barometric pressure from 30 to 26-28 because water vapor in lighter than the Oxygen (21%) and Nitrogen (78%) molecules it replaces. A typical high-pressure front is colder air with very little water vapor when compared to a low-pressure front. The barometric pressure of a high-pressure front is near or exactly 30. The water vapor is converted to rain water when these two fronts engage each other to refill all the collection points for fresh water and with the process of photosynthesis, carbon dioxide and water are chemically united to produce our only natural source of Oxygen and hydrocarbons consisting of Carbon and Hydrogen for food, vegetation, and energy products. It's God's creation that enables life to continue here on earth. I think the advocates have "earth is dying" with the "earth is very much alive".

I believe the "earth is dying" faction forgets that Carbon Dioxide remains relatively constant at 392-400 parts per million or .0392% -.04% of the atmosphere because the vegetation here on earth has a voracious appetite for Carbon Dioxide. The danger here is if the Carbon Diet and Footprint faction actually succeeds in reducing the Carbon Dioxide levels in the atmosphere worldwide and it significantly reduces the amount of food, vegetation, and energy and Oxygen creation, then the movement would cause a level of the "earth is dying". In reality, the actions of the Carbon Diet and Footprint and the "earth is dying" advocates is insignificant compared to the size and volume of the gases of the atmosphere that is so dense it presses on all humans and roof tops at 14.7 lbs. per square inch. I don't think you can make the position "earth is dying" based on simple chemistry and physics. Actually, the earth is full of life.

On a smaller scale, the earth is comprised of a finite number of atoms, and the only atom that can escape the earth's gravitation pull is Hydrogen. Hydrogen exists in the atmosphere at less than 1 PPM (Parts Per Million).

We can't change the number of atoms on earth as we are stuck with the current count. As you know, atoms switch positions often and move from one compound to another. For example, Hydrogen occurs in both water and gasoline, and my papers shows that you get 1 gallon of liquid water from each gallon of gas burned. Amazing!

Joe Schramek

Letter to Nolan Finley on Detroit News Article-February 4, 2014 Faith leaders must speak out on climate change-Charles Morris of Madonna Univ

I find it absurd that the Religious, such as the Rev. Charles Morris of Madonna University, state it is a moral responsibility to abide by the EPA and Government's Policy on Climate Change. He and the Catholic Bishop's Conference are aligning with the basic Liberal positions of the current administration while about 50% of the American citizens have a differing point of view. I would appreciate if the self-appointed religious would spend most of their time on shepherding their flock on matters of faith and morals, rather than taking sides with special interest parties who try to interfere with God's work and worry.

About a year ago, I was attending Divine Child Sunday Mass, when the Epistle reading was Matthew Chapter 6, verses 24-34. It clearly shows that worry about having enough food, drink, vegetation, flowers, clothing, and life is God's worry. I just don't see why we need Stewards of the

Earth to attempt to change what God has created, and take the worry away from Him. Here is the Matthew reading:

HCSB-Matthew Chapter 6-Verses 24-34

(Scripture quotation taken from the Holman Christian Standard Bible®. Copyright © 1999, 2000, 2002, 2003 by Holman Bible Publishers. Used by permission. Holman Christian Standard Bible®, Holman CSB®, and HCSB® are federally registered trademarks of Holman Bible Publishers.)

No one can be the slave of two masters: he will either hate the first and love the second, or treat the first with respect and the second with scorn. You cannot be the slave both of God and of money.
That is why I am telling you not to worry about your life and what you are to eat, nor about your body and how you are to cloth it. Surely life means more than food, and body more than clothing! Look at the birds in the sky. They do not sow or reap or gather into barns; yet your heavenly Father feeds them. Are you not worth much more than they are?
Can any of you, for all his worrying, add one single cubit to his span of life? And why worry about clothing?
Think of the flowers growing in the fields; they never have to work or spin; yet I assure you that not even Solomon in all his regalia was robed like one of these. Now if that is how God clothes the grass in the field which is there today and thrown into a furnace tomorrow, will he not much more look after you, you men of little faith?

Joseph E. Schramek

So do not worry; do not say, "What are we to eat? What are we to drink? How are to be clothed?" It is pagans who set their hearts on all these things. Your heavenly Father knows you need them all.
Set your hearts on his kingdom first, and on his righteousness, and all these other things will be given to you as well.
So do not worry about tomorrow: tomorrow will take care of itself. Each day has enough trouble of its own.

Contrary to the stewardship of the religious and faith leaders to speak out on climate change, the Pope was planning an encyclical on climate change. Initially, it was due out in April. Now, conjecture has it coming out in October. Hopefully, the Pope will reconsider based on his responsibility to speak with infallibility on matters of faith and morals and nothing else. However, world events could change that. Thanks for the opportunity to challenge Rev. Charles Morris's positions.

Joe Schramek

Response to Hackett High School on New Solar Panels-Goes Greener

Dear Hackett Administration and Science and Math Teachers:

I was moved by the contribution of the family of an alumni to honor their lost loved one by contributing an energy conversion system for saving electrical costs for Hackett High School. I sincerely hope that it will help lower overall costs for families to continue the fine tradition of a Catholic education for Kalamazoo Catholic families.

In 1957, I was one of the first students to use the new Chemistry class room at the eastern end of the upper floor at the St. Augustine School building and my teacher was Sister Clarita. I enjoyed science and math a lot and I went to Western where I was honored as one of the six highest ranked Freshmen Chemistry. After two years at Western, I transferred to Michigan State University where I studied and received a BS Degree in Electrical Engineering.

Joseph E. Schramek

With due respect to the donor's family, I found fault with the statements in the Hackett's Irish Pride magazine article, "Additionally, it has avoided the release of more than 10,000 lbs. of Carbon Dioxide into the atmosphere......and later, "Also, it demonstrates Hackett's real commitment to environmental stewardship and the necessity of renewable energy to confront climate change." My disagreement and positions are based on my study and interest in physics, math, and chemistry that I first learned at Catholic High Schools at Lansing St. Mary's for 2 and 1/2 years and at Kalamazoo St. Augustine 1 and 1/2 years and the State Colleges: Western Michigan University for two years, and Michigan State University for 2 and 1/2 years. Below are the points that support my positions:

1.) The absolute number of Carbon, Hydrogen and Oxygen atoms is fixed count on Earth. The only atom that escapes from the Earth's gravitational force is the Hydrogen molecule which has two Hydrogen atoms. Hydrogen is the lightest atom and the most reactive and is rare as a gas at .000055% of the total atmosphere.
2.) The amount of Carbon Dioxide is likewise rare in the atmosphere, since the amount is relatively stable at 392 parts per million or .0392% of the atmosphere. Why doesn't the percentage of Carbon Dioxide increase with the burning of gas and coal and the exhalation of humans and animals? Carbon Dioxide is the sources of generated Oxygen through the photosynthesis process of all vegetation and plant life on earth. Photosynthesis marries water, Carbon Dioxide and other trace elements together to form hydrocarbons (Hydrogen, and Carbon

Fallacy of the Green Movement and Climate Change

compounds) with Oxygen as a by-product. The 10,000 lbs. of Carbon Dioxide quoted above has a short life here on earth as all plant life has a voracious appetite for the carbon and oxygen in Carbon Dioxide and this keeps the percentage of Carbon Dioxide at 392 parts per million with an annual range of +.0003% in winter months and -.0012% at the end of the growing season. As a result, all inhabitants of Earth can rely in the percentage of Oxygen remaining relatively constant at 21% and easy to breath.

3.) The mention that Carbon Dioxide has an effect on climate change is false. There are only two significant greenhouse gases, water vapor and Carbon Dioxide. Water vapor is the more potent of the two greenhouse gases. Water vapor can reach a high of 4% of the atmosphere over the ocean, while carbon dioxide is only .0392% of the atmosphere with an annual range of +.0003% in winter months and -.0012% at the end of the growing season. The water vapor molecule had a lighter molecular weight (18) but higher amounts water vapor in the atmosphere (up to 40,000 ppm (parts per million) for water vapor versus a relatively constant 392 ppm for carbon dioxide-molecular weight (44), and has a much larger impact (100 times) on the earth's atmospheric pressure. Since the amount of carbon dioxide is almost constant at 392 ppm, and the water vapor molecule varies from 0-4% of the atmosphere, water vapor molecule is totally responsible for our violent hurricanes and tornadoes and storm damages from high-straight line winds in storms. In high pressure systems (Barometric reading of 29-30 inches of Mercury), we experience good weather with

Joseph E. Schramek

few clouds and fresh air with the highest percentage of oxygen in the air. In low atmospheric pressure systems (Barometric reading of 26-28 inches of Mercury) the air is very warm and humid with reduced oxygen in the atmosphere. Low pressure systems are sometimes incorrectly labeled heavy air. The low-pressure systems provide the water molecule which produces rain when the system is confronted with a colder temperature high pressure front that generally comes from the Northwest. These fronts produce the most devastating tornadoes and thunderstorms with high straight-line winds.

I'm so sorry that I have to confront the fine work and presentation of the Irish Pride magazine, November 2011 issue. I have great respect for the family in their contribution that will surely reduce Hackett's electrical costs. I encourage all of Hackett's Alumni to make these contributions where possible.

Best regards,
Joseph E. Schramek

Response to Tim Powers Editor of the Dearborn Press and Guide

Subject: Response Letter to Dearborn Press and Guide Article by Katie Hetrick on "Adopt-A-Watt Brings Light to Dearborn".

The article centers on the City of Dearborn using BDT Light Sources induction type lamps that feature an electrical induction circuit that replaces internal electrodes that are used in conventional fluorescent type lamps, eg., fluorescent, metal halide, and high-pressure sodium lamps that are currently used for gas stations, convenience stores, shopping malls, warehouses, sport venues, parking lots and municipal lighting. The supplier, BDT Light Sources of Livonia, Michigan claims the lights last 100,000 hours about 5-13 times longer and save 40-70% electricity over conventional type lamps. Apparently, these new lamps are more expensive than the conventional lamps and take 12-14 months to produce enough savings to warrant the purchase of these new induction type lamps.

Joseph E. Schramek

Also in the article, the "Adopt-A-Watt" program is where activists for producing less dependence on fossil fuels have already sponsored 66 of these new lamps for the city. Based on the expected savings in electrical cost savings for these new lamps, the city is planning on installing electric vehicle charging stations at various new parking structures across the city. The city shelved plans for meters to charge electric vehicle users because the estimated cost for charging an electric vehicle is only $.81/day. In other words, any expected electrical cost savings for the City of Dearborn for the new induction lighting will be given away to citizens who can afford the more expensive electric vehicle over the conventional gas powered vehicles.

My concern is that why does the City of Dearborn have the right to provide free electricity for electric vehicle owners in the future and as a result eliminate the ability to pay out the new lamps in 12-14 months. The City of Dearborn will be using tax payer's money to pay these owners for their needed daily electrical charging of their propulsion systems.

Joe Schramek

Gov. Granholm's State Address-Coal Power Plants Spew Dirty Carbon

Dear Governor Granholm,

On your STATE of the STATE address last night, you made the comment that Coal-Powered Power Plants spew dirty carbon. You did not say carbon dioxide, just carbon. I wonder why you mention that carbon is dirty. I was taught that carbon and carbon dioxide are the building blocks of life on earth. For one point, carbon is never dirty. Carbon is clear and translucent as in the form of Diamonds. As a carbon black or graphite or coke, it produces the hottest of fires as used in the manufacturing of iron ore into cast iron and steel. Carbon is the key element of wood, oil, coal, and its gases: carbon monoxide and carbon dioxide. Many environmentalists call carbon dioxide a dirty gas, but in error. Carbon dioxide along with water are the key compounds involved in the manufacture of oxygen and plant life for earth inhabitants through the process of photosynthesis. Here is an excerpt from the WEBSITE-Wikipedia:

Photosynthesis[α] is a metabolic pathway that converts carbon dioxide into organic compounds, especially sugars, using the energy from sunlight.[1] Photosynthesis occurs in plants, algae, and many species of Bacteria, but not in Archaea. Photosynthetic organisms are called *photoautotrophs*, but not all organisms that use light as a source of energy carry out photosynthesis, since *photoheterotrophs* use organic compounds, rather than carbon dioxide, as a source of carbon.[2] In plants, algae and cyanobacteria photosynthesis uses carbon dioxide and water, releasing oxygen as a waste product. Photosynthesis is crucially important for life on Earth, since as well as it maintaining the normal level of oxygen in the atmosphere, nearly all life either depends on it directly as a source of energy, or indirectly as the ultimate source of the energy in their food.[2][β] The amount of energy trapped by photosynthesis is immense, approximately 100 terawatts per year:[3] which is about seven times larger than the yearly power consumption of human civilization.[4] In all, photosynthetic organisms convert around 10,000,000,000 tonnes of carbon into biomass per year.[5]

I don't think your staff and you have thought out the total pluses and minuses relative to using electric vehicles and batteries. You may want to consider the use of electric vehicles that made the City of Detroit the number one leader in public transportation for the world in the Mid-1930's. It was done with electric buses and trolleys that picked the electric power from transmission lines above the street. Woodmere cemetery actually used an electric trolley-hearse for the transportation of caskets and family members and friends to the cemetery from Detroit Funeral homes. Actually, the major highways could be electrified and cars would have electric usage meters similar to the ones we use in our homes. Batteries are heavy and should be eliminated for a more efficient process. Vehicles could use gasoline engines to move off-line from the highways and streets that have electrical transmission lines.

Sorry, my education at Western Michigan University as a Math and Chemistry-Major and an Electrical Engineering BS from MSU does not support the changes you are pursuing for Michigan relative to issues in your address.

Joseph E. Schramek

Letter to Detroit News-Jim Nash's Letter, "The Problem with Fracking"

Letters to the Editor,

Jim Nash's letter, "The problem with fracking" is 100% wrong. He claims that 5 to 23 million gallons of freshwater per well are lost during the deep well oil drilling with fracking. Mr. Nash forgets that the oil and gas captured produces an equivalent of water plus copious amount of carbon dioxide through the burning process of newly refined hydrocarbons in autos, heating of homes, and operation of plant equipment and machines, i.e., one gallon of water and a large amount of carbon dioxide-22 lbs. The newly created carbon dioxide gas and water vapor is utilized for plant and vegetation growth that in the end produces our only natural source of oxygen for the continuation of animal and human life here on earth thru the process of photosynthesis. Magically, the earth keeps the level of carbon dioxide at the non-threatening level of 392 parts per million contrary to the environmentalist's scare tactics that the level of carbon dioxide increases and causes climate change or global warming. The carbon dioxide created by burning hydrocarbons is consumed quickly by the voracious appetite that plants and vegetation have for this gas. Also, the only way we can get energy burning hydrocarbons comes from the newly created and existing supply of the earth's supply of oxygen which magically is maintained at the level of 21% of the atmosphere.

Respectfully,

Joe Schramek

Detroit News Editors-"Michigan needs a 'no-regrets' strategy"

Jonathan Wolman, Editor and Publisher of the Detroit News
Nolan Finley, Editorial Page Editor

Dear Jonathan and Nolan,

I read the article, "Michigan needs a 'no-regrets' energy strategy", by Rebecca Stanfield and Lisa Wosniak. The theme of their paper stills worries about the effect of carbon dioxide which I believe they think is a threat to life here on earth, while thinking that the minimal reductions are the goal of the energy strategy.

Regardless of the conservative or liberal positions on the subject, carbon dioxide is the magical gas that creates all vegetation on earth and the vegetation is our only source of oxygen that all life needs for existence on this planet. There are two significant greenhouse gases, water vapor and carbon dioxide. Water vapor is lighter than all atmospheric gases and it rises until it forms clouds and eventually dew or rain water. Carbon Dioxide is much heavier than all the atmospheric gases and it lies close to the ground eventually until it mixes with water in the form of rain or dew and enters the ground to eaten by plant life to form growth of the plant. In doing so, it forms fiber and fruits of plants while giving off our pure oxygen. Amazingly, the amount of carbon dioxide never gets higher than 392 parts per million on the average and never causes changes in atmospheric pressure. Water vapor is the exact opposite, it increases in volume and can reach only 40,000 parts per million or 4% of the total atmosphere and can drop the atmospheric pressure to 26-27 inches of Mercury in hurricane and tornado conditions.

My paper attached, "The Green and Carbon Footprint Movements in the World-Why?", is probably the most accurate assessment of the fallacy created by "pseudo" environmentalists and scientists that seems to be believed by nearly all unfortunately. The "doom and gloom" they forecast has no basis in science. My paper is based on simple chemistry, math, and physics that is taught in initial level engineering and scientific studies at all Michigan universities.

My request is that you take the time to read my editorial opinion, and consider publishing it in one of your future editions.

Joseph E. Schramek

Letter to Detroit News-Artic Study Warns of Melt and 5 Foot Rise in Ocean

Subject: Nation and World-Arctic study warns of melt- Associated Press Article by Karl Ritter and Charles J. Hanley

Scientific panel predicts a 5 feet raise in sea levels by end of century from melting of Greenland, and the polar ice caps from accelerating global warming. This is impossible! The amount of water in its many forms is a fixed amount on the planet Earth. Most fear mongers mention the melting of polar ice cap adds to the ocean sea level. The fundamental error is that an iceberg is .9 the density of water, and therefore it floats with .9 of its volume below the ocean level and .1 of its volume above sea level. When an iceberg melts its density is equal to water. Therefore, when the iceberg melts it merely replaces the submerge level of the iceberg and since its density in the melted form is equal to adjacent ocean sea level, and there is no rise in the overall level of the ocean. The scientists believe the acceleration of global warming is attributable the bogged down efforts to reduce carbon dioxide emissions over the past two decades.

Most scientists overlook the importance of carbon dioxide in sustaining life on earth. Carbon dioxide is the magical gas that creates our oxygen from the process of photosynthesis. Photosynthesis marries carbon dioxide, water, and other nutrients in our soil to produce vegetation, trees and oxygen. The plant-life on Earth has a veracious appetite for carbon dioxide and this keeps the level of this gas at non-threatening levels for humans and adds the necessary oxygen levels to support human life. Fortunately, carbon is the most prevalent atom in support of our life. It's the principle atom for the manufacture of sugar, wood, cloth, plastic, coal, coke, charcoal, diamonds, carbohydrates, gas, hydrocarbons, etc.

Also, the principle leaders of the global warming have developed a situation where people are designated as believers in it or non-believers. If a situation is relegated to one where a person either believes or doesn't believe, it questions the accuracy of either argument. Since neither side is a majority, I conclude that the situation is: true, not-true, or not a significant effect. My basic and elementary statements above support a situation where the total Earth systems in place have the capability of handling the ranges of current variations of climate, and absolute levels of oxygen, hydrogen, carbon atoms that the Earth systems contain without any threats to our existence.

Joseph E. Schramek

Thank God, the Earth spins around sun at 67,000 mph and rotates at 1000 mph at the equator. Amazingly, the sun only uses the most insignificant amount of its radiant energy to satisfy the needs of Earth's inhabitants until the end of time.

It's so simple. Why can't scientists or climatologists comprehend it?

Joe Schramek

Letter to MSNBC-Chris Jansing's Show Today-Gore's 20 Feet Rise in Ocean

Dear MSNBC (Jansing's Show Today)

Fear reports over the weekend and Monday that Gore's projection is 20 feet rise in ocean level, 1 foot rise in the new documentary that supports global warning but contradicts Gore's estimate, and today, a contributor mentioned that the ocean would rise 3 feet and cause levees to be built at major US cities on the Atlantic and Gulf Coast lines. All the reports are untrue. Liquid water, ice bergs, and humidity are all the molecule H_2O or 2 hydrogen atoms and one oxygen atom for a total molecular weight of 18. The amount of water in its many forms is a fixed amount on the planet earth. Most fear mongers mention that the melting of the polar ice cap add to the ocean sea level from warmer temperatures from the increased carbon content of the atmosphere. The fundamental error is that an iceberg is .9 the density of water and therefore it floats with .9 of its volume below the ocean level and .1 of its volume above the sea level. When an iceberg melts its density is equal to water. Therefore, when the iceberg melts it merely replaces the submerge level of the iceberg and since its density in the melted form is equal to adjacent ocean sea level and there is no rise in the overall level of the ocean.

Additionally, most scientists overlook that Carbon Dioxide is the magic gas that creates our oxygen when the process of photosynthesis marries carbon dioxide, water, and other nutrients in our soil to make vegetation, trees and oxygen. The plant-life on earth has a veracious appetite for carbon dioxide and this keeps the level of this gas at non-threatening levels for humans and adds the necessary oxygen levels to support human life. Fortunately, carbon is the most prevalent atom in support of our life. It's the principle atom for the manufacture of sugar, wood, cloth, plastic, coal, coke, charcoal, diamonds, carbohydrates, gas, hydrocarbons, etc.

Also, the principle leaders of the global warming have developed a situation where people are designated as believers in it or non-believers. If a situation is relegated to one where a person either believes or doesn't believe, it questions the accuracy of either argument. Since neither side has a great majority, I conclude that the situation is either true, not-true, or not a significant effect. My basic and elementary statements above support a situation where the total earth systems in place have the capability of handle the ranges of current variations of climate, and absolute levels of oxygen, hydrogen, carbon atoms that the earth systems contain without any threats to our existence.

Joe Schramek

Responses to Recent Articles in Detroit News October, 2014 "Buyers slow to connect with electric vehicles", and "GM Chief: Time to Get Candid"

Buyers slow to connect with electric vehicles

There are only two reasons to buy electric vehicles. Your education lacked a good background in math, chemistry, physics, and energy transfers and sources, and you believe anything supported by the government actions made by people who have the same lack of education that you have. You were quick to buy the electric vehicle because you got large rebates paid by the majority of the public that is forced to support your lack of education and purchase options. Because most electric energy for these vehicles will be produced by burning coal or hydrocarbon alternative gases to generate the source of the electricity at the power outlets for recharging your batteries, and since this conversion is less efficient than burning it in the engine of your car, you will create more of the pollutant, carbon dioxide, than if you used gasoline or a hydrocarbon alternative as a sole source of energy. Oh! Carbon dioxide is the wonder gas that produces all the food supplies and oxygen generation for human life here on earth through the process of photosynthesis to create vegetation/food, oxygen and water vapor. Oh! I wonder when the EPA will awaken to make rules on the mountains of electric batteries accompany a successful propagation of the electric vehicle. The EPA is keeping quiet because this will require a huge increase in the funding for the agency for new employees and more control over the citizens of our country.

GM Chief: Time to Get Candid

GM becomes the first public business in the country to turn its organization over to a single employee, Mary Barra. Mary throws out the teamwork approach that apparently existed in this company since its inception around 1911 or so. Apparently, Mary Barra feels confident since Michael Milliken and she have generously awarded large settlements to decedent's families for the ignition switch issue where it was determined that the ignition key magically moved from the RUN position to the OFF position prior to the accident and where the air bags in the front failed to inflate and the driver and front and rear seat passengers suffered injury of death. As I reported in earlier letters to you, the brakes still work for a number of applications and at highway speeds the steering reverts to manual versus power and all people can steer the vehicle to change lanes and leave the roadway safely. One reason that Mary and Michael may have grossly overstated GM's responsibility in that the key point for getting approval for the settlement is whether the ignition was found in the RUN or OFF position. I always thought the first professionals to the scene of an accident where serious injury or death occurs, they first turn off the ignition key to prevent any danger from a live electrical system from a spark or an ensuing fire. You can't asked people who have experienced a serious head or body injury or the fatalities whether the ignition was in the RUN or OFF or ACC position prior to the accident. It is doubtful if this would be written or spelled out with any accuracy in the formal report of the accident. Certainly, it was added after the accidents in most cases when GM and Milliken publically stated the ground rules for a settlement. What GM never told the public is that anyone can restart the vehicle in less than 1-2 seconds by sliding the transmission shift lever to the N-Neutral position and turn the key to start the engine instantly.

I've given you lots of information that those people who died in the accidents related to this issue are not always innocent. They were unbelted, drunk, no license, arguments with their lovers, speeding, driving late at night, driving over 70 mph on a neighborhood cul de sac, and vehicles rolled over where air bags can't inflate due to the lack of frontal G-Forces that trigger the devices. In most cases, GM should have refused settlements where clearly the injuries and deaths are the fault of the driver.

Response to Detroit News Article on January 1, 2015-"Recall woes follow GM into New Year"

Recently, it was reported that the number of people killed in auto accidents dropped from 37,000 in 2013 to 32,000 for 2014 reporting. Your article states that the ignition switch issue has been blamed for 42 deaths and 58 injuries, a mere portion of the total count of 32,000 deaths. In my previous reporting for me, I showed horrific accident photos of damaged vehicles of some of the victims. In my opinion, there is little doubt that the proper inflation of the air bags would not have saved these victims. It is a well-known fact that air bags provide little protection for extreme impact speeds of the vehicles. It boils down to the fact that Government and the Media (including the News and Free Press) have done little to question the culpability of GM as a cause of the deaths of these victims.

I can't find the two letters in 1980 or so that I received from Joan Claybrook-Head of the National Highway Traffic Safety Administration on the proposed rulemaking that resulted in forcing Car Manufacturers to provide Explosives and Air Bags on the steering wheel and passenger side instrument panel to lessen head and body injuries on the driver and front seat occupants for frontal collisions. At the time, it was felt that these measures would reduce injury and deaths for impact speeds of 30 mph or less. It did not provide protection against head and body injuries less than death for speeds above 30 mph, nor rear end, side, rollover, and corner impacts. Impacts with vehicles with heavier weight, and higher speeds for the other vehicle for frontal impacts provided lesser protection for the occupants. My two letters were hand written on the backside of vehicle weight reports from the Ford where I was employed as an engineer in the Chassis Engineering organization. I never agreed with Ms. Claybrook's positions but I was impressed that she wrote responses to my letters.

I'm currently reading, "42 A Portrait of My Father" by George W. Bush, his son. Throughout the book, 41 Bush wrote personal hand-written responses back to all people who supported and disagreed with his positions. He did this on Saturday's. He like Obama, wrote personal letters to opposing leaders of other countries, including Chinese, Iraq, and Russian Leaders on his suggest points of actions they should take. As a result, President Gorbachev dissolve the USSR during his term as President of the US.

I did not support the implementation of air bags for the following reasons:

1.) I called it stone-age technology. You impact the occupant's body and face with a force equal and opposite to the force created by body and head traveling the same speed of the vehicle and stopping the head and body abruptly at the apex of the bags pressure before it releases the trapped hot air mass in the bag. This means the brain mass moves forward to the front skull area and the brain mass stops from 30 mph to zero in a few inches. Testing showed that sensors on the head of crash dummies that met the 1,000 HIC-Head Injury Criteria index were thought to meet the survivability level for occupants. For all cases where the vehicle hits a stationary tree or cement stationary abutment or exactly hitting another vehicle head on in a frontal impact where the impacting vehicle has the same speed, and the same vehicle weight, the two conditions would have the same effects on the head and body of the occupants. Any speeds above 30 mph, the occupants would have greater chances for very serious injury of fatality regardless of the specific conditions of the impacts and vehicle speeds.

2.) Putting two explosive charges, one for the driver and one for the passenger in every vehicle manufactured did not make any sense to me when explosives can't be sold on the shelves in auto parts stores, supermarkets, sporting shops, or hardware stores, yet we will place them 12-16 inches from the driver and front seat occupants. Misfiring of the air bag could injure and kill occupants who are unbelted and close to any explosives that misfire.

3.) Most drivers and occupants will never use their air bags, i.e., have an accident where the air bags would not inflate, or never have an accident at all. I have not had in accident in over 55 years and I have had them in about 60-75 vehicles that I have leased as a Ford employee or purchased for my family. I have never used my seat belts also, but I always wear them. You only use your seat belts when you have an accident.

4.) Car manufacturers love the Government and the NHTSA as they add content to autos and trucks. This enables them to raise car prices and make more profit from the markup of safety and regulated items mandated by the Government.

Ms. Claybrook suggested that I become a regular approved commentator on future rulemaking by the NHTSA. Being a Ford employee, I couldn't formally comment or take positions on these issues.

As I mentioned in an earlier letter, I feel GM took the wrong route in not fighting those who had questionable accidents situations with their children who lost their lives by driving recklessly at high speeds and poor sobriety. I think the News Media could have told the other side of story and questioned the merits of the actions taken. GM may have a hard time surviving if the economy deteriorates.

Likewise, the air bag issues with Takata will be a huge waste of repair costs for Auto Producers that used these products. Compared to the 32,000 lives lost for other reasons, it doesn't make sense for the producers to be held responsible for Takata air bags past regular warranty limitations. The producers are not responsible for costs on other vehicle parts and components that are out of warranty.

Joseph E. Schramek

Let the owners pay for new air bags from Takata if the cars are out of warranty.

I wish you would expose the Government and the NHTSA for their purportedly safety devices. With 32,000 traffic deaths each year, it is grossly wrong for anyone to say a vehicle is safe to drive. Cars are like airplanes, boats, buses, trains, space capsules, bicycles, motorcycles, and any other device to move from Spot A to Spot B. Loss of life can occur using any of these forms of transportation. None of them are safer if you lose your life using them.

I would hope you could deepen your investigative journalism to switch the cause of these issues back to the NHTSA for their adoption of safety devices and rules that have little or no effect on the survival rate for the public from using transportation products when parts of components fail in use.

Joe Schramek

Dearborn Press and Guide Article Published February 26, 2012- Your Point of View-"Support for new Goodwill store at West Dearborn Location

This article shows the author's early life in the City of Dearborn, Michigan when I came here in 1961 to begin a 44-year career as an engineer for the Ford Motor Company. It shows how the city changes over time and that new business have to overcome skeptics to succeed, sometimes.

To the Editor

I've been a resident of Dearborn, since 1961. I first rented a room at 22434 Garrison owned by the Jakabowski Family of Inkster. The mother, Helen Jakabowski, had two daughters who worked for U.S. Congressman, John Dingell, one in his Dearborn office and the other in his Washington D.C. office. The home was the former residence of the first Mayor of Dearborn, Clyde M. Ford. He also owned the first Ford Dealership in West Dearborn.

I was very familiar with the proposed location of the new Goodwill Industry Store at the old Inca Computer

Joseph E. Schramek

Store and prior to that, the Pier I Imports Store. In 1961, I bought all my food from the Wrigley Grocery Store, got my haircut at the barber shop (eventually Michael's Barber Shop), across from the Mercury Coffee Shop, owned by Gus, where I ate evening meals and breakfasts on weekends. I was a charter member of the new Vic Tanny gym, where I lost 30 pounds of pizza fat.

The point of discussing all the spots in West Dearborn is that none of them exist anymore. Life is full of changes, and I see nothing wrong with a nonprofit organization opening a second-hand store in West Dearborn at the former Inca site.

My cousin, Robert Manning of Chesaning, who has cerebral palsy, got a responsible job for Goodwill Industries at one of its locations in Saginaw.

Deep said these stores would lower the business values in the immediate area and he mentions that a mortgage company complained that the business would lower its bank holdings. Strange that a mortgage company is held in such high esteem and as a quality business in light of the most prevalent loss of confidence in America that rests with these types of operations.

My point is, that over the 51 years I have lived in the Dearborn community, all the businesses go through change and eventual dissolution. I don't see why the Dearborn Planning Commission in punishing the Goodwill Industries, an American nameplate. They have spotted an opportunity to serve the city residents with lower cost products, clothing, books, and other useful products.

My wife and I visited the Salvation Army main operation in Romulus every Wednesday to make our investments in merchandise that is being sold at under value with name brand quality.

As an example, last year I bought five name brand autographed putters for less than $10 total, saving over $1,000.

I'm sure we would be one of their first customers at the new Goodwill Industries store if only the Dearborn Planning Commission and Said Deep would change his vote to an Aye. I'd like to revisit the community where my wife and I were married at Sacred Heart Church at Michigan and Military.

Joe Schramek
Dearborn Heights

Detroit News-HMO Versus BCBS Costs-March 6, 1979

In 1979, HMO's came into existence and they offered health care at lower costs by giving free annual physicals to children and adults, but you had to wait your turn at HMO facilities and used the doctors on service at the time. Today, HMO's operate similar to the BCBS system where you could keep your doctors and have regular office calls at physicians of your choice from the approved doctors in the system, the same as BCBS did for years. My letter that was published showed why costs were lower and the reasons why.

HMO — LOW COST health or lack of health care? HMO produces lower hospital bed usage than Blue Cross-Blue Shield (465 days per 1,000 subscribers for HMOs; 856 for the Blues). Sounds like preventive medicine and health care is the answer to all our problems.

But wait. Did you consider that HMO benefits young families with children and that HMO may have more youthful subscribers than the Blues? We all know that older people have longer hospital stays than young people. That may be the answer to the difference between the two health plans in terms of hospitalization.

Why hasn't somebody properly weighed the data on hospital bed usage and then compared the health plans?

How many of us have, like myself, waited four hours for health care in an HMO facility? Anyone in business knows that you can lower the cost of a service by making that service harder to obtain. But the lower costs involve a form of Russian Roulette health care.

Let's not buy a pig in a poke. Let's consider all the facts about HMO, from every angle, before we make a decision.

J. E. SCHRAMEK
Dearborn Heights

"A Christmas Favor", By My Mother, Gertrude Schramek

The actual story occurred in December, 1953, when my mother entertained 3 foreign students who assembled at Michigan State University at a conference for Foreign Exchange Students Program over the Christmas Holidays. My mother shared our Christmas celebrations with different students each year for 3 or 4 years until our family moved to Kalamazoo in 1956 when my father started a new job there.

A Christmas Favor

**by Gertrude Schramek
as told to
Pat Judge**

It was December 20 already. I noticed with alarm. Arms full of ribbons and wrapping paper, I looked at our oversized wall calendar with a sigh. Only four more shopping days 'til Christmas and I had the wrapping to do, candy to make with the children, the house to clean, and . .

The ring of the phone interrupted my thoughts. What now? I thought as I picked up the receiver. "Hello, Gert, is that you?" my friend Jean ventured.

"Yes, it's me," I said, "just a little out of sorts and wondering how I'll get 10 day's work done in four."

Jean was chairman of the Christmas in World Understanding program. I knew she was far busier than I, trying to find homes for visiting exchange students for the Christmas holidays. I cut the recitation of my own problems short and asked her how the project was going.

"Fine, a record number of students in the program but that's caused another problem — finding enough homes." She hesitated. "I know how busy you must be but I'm afraid I'm going to have to ask a favor of you."

I knew what was coming. "You want us to take a student for the holidays," I replied. "I'd like to, Jean but I can't this year. One of the boys is bringing some fraternity brothers home and it would be just too hectic. I have a list a mile long and not a thing accomplished yet. But please try me next year."

Back at the kitchen table, I finished wrapping the first batch of presents and, with satisfaction, put the first checkmark on my "to be done" list. The calendar again caught my eye and I thought of

Joseph E. Schramek

Jean trying to find a home for a young person who needed some warmth and love at Christmas. She had only four days, too.

Suddenly, my list seemed insignificant. Jean was doing something which symbolized true Christmas spirit while I puttered around with my lists and petty problems. On impulse I called her back and agreed to welcome a student into our home.

Christmas Eve, we greeted a lovely Filipino girl named Victoria. She exclaimed, "I am so happy. This is my first American Christmas" but her smile was sad. This was her first Christmas away from her family and our merry bunch must have reminded her of Christmases past. She was aching with loneliness for her family which was celebrating the birth of the Savior some 10,000 miles away.

As we walked the short distance to our church for midnight Mass, I enjoyed the beautiful crisp air and the powdery snow which swished aside us we walked. Victoria, however, pulled the collar of her new coat closer and shivered. She was not used to this weather.

During the service, Victoria sat beside me, her eyes closed. I sensed she was literally thousands of miles away from us. She heard the same "Gloria in Excelsis Deo" and saw the same Nativity scene she would at home but she could not find a familiar face in this entire congregation of 800. Then, I saw the tears she couldn't hold back any longer.

"Dear God," I prayed. "I thank you for our family and friends, gathered here to welcome Your Son. Please help Victoria to feel at home with us. Please work through us to make her feel loved and at one with her family, even though they are so far away from her tonight."

After the final chorus of Christmas singing ended, we greeted our friends, among them the Hansons who invited us for Christmas Day breakfast. When we introduced Victoria, Tom Hanson remarked: "It must be difficult for you to be away from your loved ones and your beautiful country during Christmas. I spent the holiday season there during the war and if it weren't for a family that invited me into their home, I probably couldn't have stood my loneliness."

As we made our way down the steps, we finalized plans for the next morning. Tom continued, "When you come over, Victoria, I'll get out the movies that I took of my friends in the Philippines. Would you like that?"

Victoria came to life for the first time that evening. "That would be wonderful," she said. "To see my country again so soon, just think of it!"

"Believe me, I know how you feel. I don't know what I would have done without the kindness of the Rufino family."

Victoria looked startled. "You don't mean the Rufinos from Mellos?"

"That's right," Tom said eager-

I would have missed the chance to participate in His plan

ly. "Do you know them?"

"I can hardly believe all this," Victoria said with obvious delight. "The Rufinos are my family's best friends. Our fathers were roommates in college. We are very close to them."

Victoria didn't get much sleep that night but I was thankful that it was due to happiness, not homesickness.

As Tom Hanson showed the movies, Victoria danced with joy, telling us excitedly all about her homeland and her family's friendship with the Rufinos. "I am so happy," she said, this time with real feeling. "This is my first American Christmas."

A remarkable coincidence? I know better. Christmas is surely a day when God looks down with special kindness on the lonely who are hurting inside, waiting with hope for His Peace to enter their hearts. If I'd followed my first inclination and refused to share our home, I would have missed the chance to participate in His plan.

As in the prayer of St. Francis of Assisi, the Lord had indeed made me an instrument of His peace. ■

Mrs. Judge resides in Oconto, Wisconsin

Top Points and Arguments Used to Validate "Fallacy of the Green Movement and Climate Change" Claims from the 1ˢᵗ Four Chapters of the Book

The tide is shifting as many are now recognizing the necessity of carbon dioxide as the miracle gas that provides us the only natural source of oxygen for human and animal life and the vegetation on earth that is created by the photosynthesis process that marries water and Carbon Dioxide into food and our energy supplies to continue the existence of human and animal life here on earth. These points and arguments substantiate the claims of the book.

Topic 1-The Green and Carbon Footprints of the Green Movement, Why?

- In high pressure systems (Barometric reading of 29-30 inches of Mercury), we experience good weather with few clouds and fresh air with the highest percentage of oxygen-21% and lowest humidity in the air for comfortable breathing. In low atmospheric pressure systems (Barometric reading of 26-28 inches of Mercury) the air is very warm and humid-3% water vapor and with reduced oxygen-about 20% in the atmosphere for poorer breathing conditions and discomfort as perspiration condenses on skin without evaporation and cooling of the skin on our body. Low pressure systems are sometimes incorrectly labeled heavy air.

- The low pressure systems provide the water molecules which produce rain when the system is confronted with a colder temperature high pressure front that generally comes from the Northwest in the Northern Hemisphere. These fronts produce the most devastating tornadoes and thunderstorms with high straight line winds. For some reason, the scientists ignore the water vapor molecule as the cause for all the climatic temperature disasters and global warming (real or unreal or insignificant).
- Because there is increasing burning of fossil fuels in the world, they incorrectly choose the other greenhouse gas, carbon dioxide as the culprit, even though the percentage of carbon dioxide in the atmosphere is constantly in the range of 392 ppm or .0392%.
- What keeps the percentage of carbon dioxide relatively constant at 392 ppm or .0392% is that plant life has a voracious appetite for the carbon dioxide and water molecule to create vegetation to cover the earth's soil and provide food supplies for animal and human life.
- A simple analysis is that the voracious appetite of plant life for carbon dioxide is matched almost equally to the creation of carbon dioxide through burning hydrocarbons for heating and energy for manufacturing processes.
- Hydrocarbons in nature include all plant and animal life, and fossil fuels (oil, and coal). Hydrocarbons are organic

compounds that are produced naturally from water and carbon dioxide through photosynthesis.
- If the current Green and Carbon Footprint Movements ever reach its desired goals for significantly reducing the carbon dioxide molecule percentage of .0392% or 392 ppm levels, humans and all animal life would begin to experience breathing issues causing many variations of health concerns including possible death. The Oxygen molecule in nature only comes from the Carbon Dioxide molecule through the photosynthesis process that produces our vegetation and plant life on earth.

Topic 2-EPA's Position Carbon Dioxide -CO_2 is a Pollutant, IS DEAD WRONG

- Water and Carbon Dioxide gases in the atmosphere are the main greenhouse gases in the atmosphere.
- Water vapor or humidity is the most prevalent greenhouse gas at a maximum of 4% of the atmosphere at sea level and carbon dioxide pollution is relatively constant at .0392% or 100 times lesser than gaseous water vapor in the atmosphere.
- The chemical equation for photosynthesis is shown below:

$$\text{Carbon Dioxide} + \text{Water} \xrightarrow[\text{Chlorophyll}]{\text{Light}} \text{Glucose} + \text{Oxygen}$$
$$6CO_2 + 6H_2O \longrightarrow C_6H_{12}O_6 + 6O_2$$

- The oxygen gas produced from the photosynthesis process is the sole source of new oxygen in the atmosphere for all human and animal life on earth.

Fallacy of the Green Movement and Climate Change

- Currently, because of the voracious appetite of plant life for the pollutant carbon dioxide, oxygen gas remains at a relative 21% of the total atmosphere on earth, thanks to the natural environmental process of photosynthesis.
- The conclusion is that the vegetation on earth consumes all of the pollutant, Carbon Dioxide.
- With the exception of Hydrogen, all the carbon and oxygen atoms that existed here on earth when it settled in orbit around the sun are still here today. Hydrogen is the only element that can achieve an escape velocity greater that the earth's gravitation force.
- Water vapor or humidity can lower atmospheric pressures from 30 mm for low humidity and fair weather and 26 mm for high humidity for tornado and hurricane weather conditions.
- Because Carbon Dioxide stays at insignificant levels in the atmosphere, it does not have any effect on weather conditions despite the EPA ruling it a pollutant.

Topic 3-Acid Rain-Sulfur Dioxide(SO_2) and Nitrogen Dioxide (NO_2) Products From Dirty Coal Burning Process

- Acid rain has a PH of 5.5 which is a weak acid. Others may argue whether the acid is predominately HCO_3 Carbonic Acid or others including: nitric, sulfurous, and sulfuric acids too.
- Sulfuric acid, H_2SO_4, cannot be created from SO_2 in the atmosphere. The EPA erroneously reports that H_2SO_4 Sulfuric Acid can be produced in the air from SO_3, but it requires the presence of the catalyst, NO_2

Nitrogen Dioxide. This has no chance or the remotest of possibilities of occurring in the atmosphere as both SO_3, and NO_2 gases are present at only .00005% of the air.
- Sulfurous acid, H_2SO_3, is simply produced by combining the gas SO_2 with water vapor molecule to form a gaseous molecule of H_2SO_3. H_2SO_3 only exists as a gas molecule and does not exist on record as a liquid solution. It reverts back to SO_2 Sulfur Dioxide gas in a liquid state.
- NO_2 Nitrogen Dioxide occurs in the dry atmosphere at .00002%. When the gas molecule is mixed in the atmosphere with water vapor the water vapor molecule, it forms one molecule of HNO_3 Nitric Acid. The acid molecule created is very sparse in the atmosphere. Its life is short because it is a strong oxidizer and oxidizes metals and vegetation. Its damage would be insignificant because of the weakness and strength of the acid overall.
- Nitrous oxide (N_2O) occurs in the dry atmosphere at .0000325%. It does not produce Nitric Acid in the atmosphere or anywhere else.
- It is known that rain water has a pH of 5.5 which is a weak acid.
- Studies indicate that acid rain does not kill trees. EPA admitted above that Acid Rain cannot kill trees alone. Then the EPA doesn't know, and is guessing that Acid Rain and along with other causes resulted in the death of trees in the forests areas located at high elevations in the Eastern United States.
- The EPA and other agencies use the following average carbon content values to estimate CO_2 emissions:

- CO_2 Emissions from a gallon of gasoline: 8,887 grams or 19.6 lbs. of CO_2 / gallon1
- CO_2 Emissions from a gallon of diesel: 10,180 grams or 22.4 lbs. of CO_2 / gallon2
- Water Vapor Emissions from a gallon of gasoline: 3,780 grams or 8.34 lbs. of H_2O / gallon.

- The irony of the amounts of Carbon Dioxide and Water produced from burning coal and gasoline is that the EPA never mentions that Carbon Dioxide is the miracle gas that produces all of our food through photosynthesis along with the water created in burning of hydrocarbons, and the only oxygen supply used by mankind on earth. They only see them as pollutants and detrimental to existence of mankind on earth, unbelievable.
- Water or water vapor averages .25% of the atmosphere, but it can raise to 4-5% of the atmosphere in drastic low pressure systems that are the chief source of hurricanes, tornadoes, cyclones, straight-line high winds, and excessive wind and flood damages as a result. Water vapor is the largest concentration of the greenhouse gases and can be 100 times more that carbon dioxide in high temperature/high humidity conditions of an extreme low pressure system involving 26-27 barometric readings. The remaining gases on the list are trace gases, including: greenhouse gases: carbon dioxide, methane, nitrous oxide, and ozone, and various natural, chemical, and industrial gases.

Joseph E. Schramek

Composition of dry atmosphere,

Gas	Volume
Nitrogen (N2)	78%
Oxygen (O2)	21%
Argon (Ar)	.9%
Carbon dioxide (CO_2)	.04%
All other trace elements and gases	.06%

Not included in above dry atmosphere:
Water vapor (H_2O) ~0.25% over full atmosphere, locally and large low pressure fronts 0.001%–5%

Topic 4-Convenient Lies-The Measurement of Carbon Dioxide at Volcano Sights-NO NEED TO DO ANYTHING TO REDUCE CARBON DIOXIDE

- Al Gore's book, "An Inconvenient Truth-THE PLANETARY EMERGENCY OF GLOBAL WARMING AND WHAT WE CAN DO ABOUT IT", uses carbon dioxide measurements taken at the NOAA-National Oceanic and Atmospheric Administration observatory located on the north side of the Mauna Loa Mountain at the Southern End of the Big Island of Hawaii at 4 miles and 2,600 feet below the summit of the mountain, 13,675 feet above sea level.
- The observatory records show an increase in the CO_2 concentration from 330 ppm (parts per million) in 1975 to 400 ppm in 2014. They continued their efforts in spite of the fact that Mauna Loa is an active volcano which erupted in 1950, 1975, and 1984, and with between minor

eruptions and venting of volcanic gases continuously and the continuous and ongoing volcanic activity a few miles to the Southeast on the Hawaii Island where carbon dioxide has been continuously emitted into the local atmosphere.
- The observatory records show an increase in the CO_2 concentration from 330 ppm (parts per million) in 1975 to 400 ppm in 2014. They continued their efforts in spite of the fact that Mauna Loa is an active volcano which erupted in 1950, 1975, and 1984, and with between minor eruptions and venting of volcanic gases continuously and the continuous and ongoing volcanic activity a few miles to the Southeast on the Hawaii Island where carbon dioxide has been continuously emitted into the local atmosphere.
- This lie, that measuring ambient CO_2 concentrations along with volcanic gaseous releases below the mountain top of an active volcano is OK, is used by Al Gore to show the key point that the level of carbon dioxide is increasing across the world, and alleges the change causes perception of climate change or global warming.
- Magma contains gases that are released into the atmosphere during eruptions. Gases are also released from magma that either remains below ground (for example, as an intrusion) or is rising toward the surface. In such cases, gases may escape continuously into the atmosphere from the soil, volcanic vents, fumaroles, and hydro-thermal systems. The most abundant gas typically released into the atmosphere from volcanic systems is water vapor (H2O), followed by carbon dioxide CO_2, and sulfur dioxide (SO_2).

Joseph E. Schramek

- Since the Mauna Loa has no vegetation from the summit and past the observatory, conversion of carbon dioxide CO_2 into vegetation and oxygen from photosynthesis does not occur there and the gas being the heaviest of the principal atmospheric gases, the carbon dioxide CO_2 would collect at low spots along the mountain surface downward and accumulate over time. This would interfere with the natural flow of the oceanic atmosphere and taint any measurement taken.
- The fact that Scientists use the Mauna Loa Observatory for the determination of the concentration of carbon dioxide CO_2 has to be a major error. To alleviate any error the observatory should have been closed and a new one erected in a non-volcanic location and not even remotely close to one.
- Al Gore's 2nd Convenient Lie in his book, "An Inconvenient Truth-THE PLANETARY EMERGENCY OF GLOBAL WARMING AND WHAT WE CAN DO ABOUT IT", is the omission of water vapor (H_2O) as the more prevalent greenhouse gas over carbon dioxide CO_2.
- The evidence presented here confirms that carbon dioxide is not a pollutant but necessary to continue life here on earth and it does not have any effect on climatic events as purported by Al Gore's book, "An Inconvenient Truth-THE PLANETARY EMERGENCY OF GLOBAL WARMING AND WHAT WE CAN DO ABOUT IT". Carbon dioxide is the most important gas in our atmosphere and environment, because it gives us the only natural source of oxygen, vegetation, and plant life for food through the process of photosynthesis to enable human and animal life here on earth to continue.

NO NEED TO DO ANYTHING TO REDUCE CARBON DIOXIDE!!

- Most fear mongers mention the melting of polar ice cap adds to the ocean sea level. The fundamental error is that an iceberg is .9 the density of water, and therefore it floats with .9 of its volume below the ocean level and .1 of its volume above sea level. When an iceberg melts its density is equal to water. Therefore, when the iceberg melts it merely replaces the submerge level of the iceberg and since its density in the melted form is equal to adjacent ocean sea level, and there is no rise in the overall level of the ocean.

Author's Statement at the Goodread's in Support of My Book

I wrote the book. My knowledge of Engineering levels of Chemistry, Physics, and Math and my life-long interest in the atmosphere gives me the insight to make accurate assessments on how dependent mankind is on the miracle gas, Carbon Dioxide, and how this gas is responsible for our natural source of oxygen and our food and energy suppliers to continue animal and human life here on earth forever. Some of my key points and arguments that support the fallacy are:

Rain or water vapor-H_2O in our atmosphere is a greenhouse gas the same as Carbon Dioxide CO_2 and in low pressure storm fronts is 100 times more prevalent, 40,000 ppm versus 392 ppm for Carbon Dioxide and it lowers the Barometric Pressure from 30 to 26-27 inches. The merger of low pressure hotter weather fronts with colder high pressure weather fronts is responsible for all severe weather events: hurricanes, tornados, high straight line winds, cyclones, typhoons, floods, and blizzards with high snow accumulations. Many involved in the Climate

Fallacy of the Green Movement and Climate Change

Change Movement use these weather extremes as a result of higher global temperatures. These impacts of global warming have no merit here, as these events are normal, and have occurred randomly with equivalent storm damage over modern history. This is repeated frequently in the articles in my book. My introduction in the Author's Note shows that I became interested in this topic based on my experience at 8:30 PM on June 8, 1953 when the storm passed through Lansing, Michigan that resulted in one of the Ten Worst Rated (F5) Tornadoes in US History at Flint-Beecher, Michigan that killed 116 people. Extreme weather events are caused by low pressure fronts, 26-27 Barometric Pressure, and not global warming. There is no evidence that global temperatures are increasing. I didn't include this topic in my book, because it can only occur with more radiant energy output from the sun or a reduction in the earth's rotational speed at the equator (one revolution every 24 hours) to a longer period of time. Since there is no evidence of this occurring, global temperatures will stay within current ranges of variability over time.

"Why is the burning of fossil fuels and hydrocarbons important for continuation of life here on earth." This process produces copious amounts of water vapor H_2O and carbon dioxide CO_2. Both of these molecules through photosynthesis produce all of our vegetation for food and energy products, and the only natural source of Oxygen O_2 supply for continuation of human and animal life here on earth. In addition, we need Oxygen O_2 for the ignition of fossil fuels and hydrocarbons to enable continuation of completing the ion exchanges for continued life here on

Joseph E. Schramek

earth.) Most people forget that the climate change believers want to reduce CO2 by reducing the use of fossil fuels. Fossil fuels produce CO_2 by consuming O_2. When the CO2 is converted by photosynthesis it replaces exactly the amount of O_2 that was consumed. If you stop burning fossil fuels, you also stop the consumption of O_2, and therefore, there is no impact on the percentage of O_2 in the atmosphere. Why it is important to maintain the available Oxygen O_2 levels in our atmosphere near 21%. Complete stoppage of burning fossil fuels would have some effect, but many natural sources of Oxygen O_2 like decay of vegetation, volcanoes, fires, human and animal respiration, industrial exhausts, etc. would be a substitute but new vegetation would be reduced due to less water vapor and carbon dioxide CO_2 available for photosynthesis.)

Melting ice that is floating in the ocean and sea water cannot raise the level of the water. This is because water expands upon freezing as ice crystals are formed and contracts when ice melts. You can prove this to yourself quite easily. Put a couple of cubes of ice in a glass jar that has straight sides. Then pour water into the jar until all the ice is floating, meaning there is no ice touching the bottom of the jar. Now mark the level of the water and let the ice melt. Though the top of the ice cubes were above the level of the water, once the ice has melted, the water will still be at the same level. The experimental glacier melted and the sea level didn't rise. I didn't mention the melting of ice flows or glacier fields on land as the water vapor in most cases will evaporate and become new clouds and never see the ocean. I guess it would be insignificant relative to the level of the ocean, as all rivers, streams, rills, creeks, lakes, ponds, and seas also end up in the ocean yet the ocean remains at normal levels as input and output to it remain in balance. The conclusion here is that melting of polar ice at the North and South poles cannot raise the level of oceans at all. Actually, the ice at both poles came from water vapor in the atmosphere from predominately oceans, seas, lakes, rivers, water deposited by rain on land, rias, streams, creeks, etal and it lowered the levels of sources of water and by melting it is replenishing it to where it came from.

My book clearly shows that all extreme weather events are normal, and random and are simply caused by the water vapor molecule in low atmospheric pressure fronts that is 100 times more prevalent than the carbon dioxide molecule.

I didn't write the book for man-caused global warming. It doesn't exist as it requires an increase in energy output from the Sun or a slowing of the rotation of the earth about the axis would lengthen the earth's daily exposure to the sun and it would increase the earth's average temperatures or global warming. My other concern is that the use of wind-mills to create electrical power could exert a reaction force opposite to the rotation of the earth and eventually begin to lessen the speed of rotation, and cause the planet to get hotter.

Author's Rebuttal to Glen R. Stott's Criticism of My Book As Recorded in Goodread's September 21, 2015

Glen R. Stott's criticism of my book will be shown in regular text and preceded by the name, **Stott**, and my rebuttal to his criticism will be shown in bold-face italicized text and preceded by the name, **Schramek**.

Stott-I believed in man-caused global warming until 2007. I subscribed to Science and Scientific America and read their articles every month. When they talked about global warming, it was always from the perspective that it existed and was caused by man. When the dire predictions for 2007 didn't materialize, I began to look at global warming with a bit more skepticism. I found the evidence of the fraud primarily in the writings of those who promoted man-caused global warming. I began to see serious flaws in their work and their conclusions. I have read very few books that argue against man-caused global warming, since global warming falls apart on its own weight, but I thought I would give Schramek a shot. I was mostly disappointed. His book consists of a number of essays and letters he has written, most of which are cut-and-paste repetitions of each other. In the 89 pages, there are about 15-20 pages of original material. Almost all of that contains incomplete arguments or non sequiturs.

He provides a very accurate description of the part H_2O plays in storms from the standpoint of meteorology. However, climatology is a completely different subject. While H_2O is the prominent player in storm development, a gradually warming planet equals gradually higher levels of H_2O in the atmosphere resulting in gradually more violent storms. Schramek's weather argument, while interesting, has no bearing on the impacts of global warming.

Schramek-Rain or water vapor-H_2O in our atmosphere is a greenhouse gas the same as Carbon Dioxide CO_2 and in low pressure storm fronts is 100 times more prevalent, 40,000 ppm versus 392 ppm for Carbon Dioxide and it lowers the Barometric Pressure from 30 to 26-27 inches. The merger of low pressure hotter weather fronts with colder high pressure weather fronts is responsible for all severe weather events: hurricanes, tornados, high straight line winds, cyclones, typhoons, floods, and blizzards with high snow accumulations. Many involved in the Climate Change Movement use these weather extremes as a result of higher global temperatures. His concern related to the impacts of global warming have no merit here, as these events are normal, and have occurred randomly with equivalent storm damage over modern history. This is repeated frequently in the articles in my book. My introduction in the Author's Note shows that I became

Joseph E. Schramek

interested in this topic based on my experience at 8:30 PM on June 8, 1953 when the storm passed through Lansing, Michigan that resulted in one of the Ten Worst Rated (F5) Tornadoes in US History at Flint-Beecher, Michigan that killed 116 people. Extreme weather events are caused by low pressure fronts, 26-27 Barometric Pressure, and not global warming. There is no evidence that global temperatures are increasing. I didn't include this topic in my book, because it can only occur with more radiant energy output from the sun or a reduction in the earth's rotational speed at the equator (one revolution every 24 hours) to a longer period of time. Since there is no evidence of this occurring, global temperatures will stay within current ranges of variability over time.

Stott-Another specious argument Schramek makes is; since the conversion of CO_2 to O_2 during photosynthesis is the major source of O_2 in the atmosphere.

Schramek-*Wrong! It is the only natural source of oxygen in our atmosphere.*

Stott: Wrong! It's a minor point, but there are other natural sources of O2 in the atmosphere. Being a knowledgeable meteorologist, I would expect you would be aware that lightening is one natural source of O2 in the atmosphere. There are others. Still, photosynthesis is, by far, the most important.

Schramek-*Correct! On Page 4, the Paragraph, "Why is the burning of fossil fuels and hydrocarbons important* *for continuation of life here on earth." This process produces copious amounts of water vapor H_2O and carbon dioxide CO_2. Both of these molecules through photosynthesis produce all of our vegetation for food and energy products, and the only natural source of Oxygen O_2 supply for continuation of human and animal life here on earth. In addition, we need Oxygen O_2 for the ignition of fossil fuels and hydrocarbons to enable continuation of completing the ion exchanges for continued life here on earth.*

Stott- He forgets that the climate change believers want to reduce CO_2 by reducing the use of fossil fuels. Fossil fuels produce CO_2 by consuming O_2. When the CO_2 is converted by photosynthesis it replaces exactly the amount of O_2 that was consumed. If you stop burning fossil fuels, you also stop the consumption of O_2, and therefore, there is no impact on the percentage of O_2 in the atmosphere. However, if the solution were to sequester CO_2 as some propose, Schramek's argument would need to be examined.

Schramek-*True! That why Page 5, "Why it is important to maintain the available Oxygen O_2 levels in our atmosphere near 21%. Complete stoppage of burning fossil fuels would have some effect, but many natural sources of Oxygen O_2 like decay of vegetation, volcanoes, fires, human and animal respiration, industrial exhausts, ..etc. would be a substitute but new vegetation would be reduced due to less water vapor and carbon dioxide CO_2 available for photosynthesis.*

Joseph E. Schramek

Stott-Regarding the need for fossil fuel to provide O_2 through photosynthesis: As stated before; Every single molecule of O_2 that comes from photosynthesis of a CO_2 molecule created by burning fossil fuel was pulled from the atmosphere by combustion of the fossil fuel. The O_2 molecule that was put in the air by photosynthesis only replaced the O_2 molecule that was taken from the air during combustion. Before man discovered fire, the CO_2/O_2 balance was fine – massive burning of fossil fuel by human beings was not needed. Why would it be needed now?

Stott-Schramek makes other less important claims that also fall apart under scrutiny. In addition, he claims that CO_2 is a constant and is not rising. It would be helpful if he had footnoted that claim. He does claim that the most important CO_2 monitoring station, which is in Hawaii, is inaccurate. His basis for that claim falls far short of being convincing. I have a different reason for doubting the Hawaii station, but it is only conjecture at this point. Rising CO_2 is not really important anyway, since there is no reliable climate model that shows a provable, quantifiable connection of global warming to CO_2.

Schramek-*On Page 29, in the Section, "Inconvenient Lies-The Measurement of Carbon Dioxide at Volcano Sites", a critique of Al Gore's Book, why would NOAA-National Oceanic and Atmospheric Administration measure Carbon Dioxide CO_2 on top of two active volcanoes where volcanoes vent fumes of water vapor and carbon dioxide continually. Other check points are in the vicinity of active volcanoes along the Pacific rims* *with similar results. The rise of 40-50 ppm over 35 years is dwarfed by the main cause of extreme weather events water vapor as mentioned before where it is 40,000 ppm compared to the diminutive level of 392 ppm for Carbon Dioxide CO_2.*

Stott-In fact, all previous warming periods show CO_2 and temperature rising roughly together, but temperatures generally have gone up first, followed by CO_2. So what is cause and what is effect? Or is it possible that both temperatures and CO_2 levels are impacted by a third variable? In the so-called modern period, temperatures started rising roughly 1000 years ago during the Medieval Warming Period – long before increasing CO_2.

Schramek-*A dizzy scenario of Who's on First? 1000 years ago!!! My book clearly shows that all extreme weather events are normal, and random and are simply caused by the water vapor molecule in low atmospheric pressure fronts that is 100 times more prevalent than the carbon dioxide molecule as mentioned repeatedly in my book.*

Stott-First, my basic argument about "all previous warming periods" is not about temps 1000 years ago but about an "ice age – interglacial" cycle that has been going for 500,000 years. Somewhere in my comments I pointed out that your argument above is perfectly correct in describing causes of individual storms, but totally ignores the arguments that Climate Change advocates make, namely, that over time, the amount of water vapor and energy in the atmosphere gradually grows because of warming. This creates more violent storms caused by gradually increased quantities of water in the atmosphere and wider disparages in pressures. The CO_2 in that atmosphere has no quantifiable impact on the extremeness of a storm. The claimed impact is a gradual warming which creates more H_2O and energy in the atmosphere and then, those (not CO_2) increase the intensity of the storms.

Stott-I give Schramek three stars on a no-star book for his water/ice argument. It never occurred to me, nor have I read anyone else who has pointed it out. Melting ice that is floating in water cannot raise the level of the water. This is because water expands upon freezing as ice crystals are formed and contracts when ice melts. You can prove this to yourself quite easily. Put a couple of cubes of ice in a glass jar that has straight sides. Then pour water into the jar until all the ice is floating, meaning there is no ice touching the bottom of the jar. Now mark the level of the water and let the ice melt. Though the top of the ice cubes were above the level of the water, once the ice has melted, the water will still be at the same level. The experimental glacier melted and the sea level didn't rise. This argument falls apart if the ice is not floating in the water. If you support the ice above the water level, when it melts, the water level will rise. Schramek doesn't mention this little problem.

Schramek-*Why don't the Climate Scientist's know this? I didn't mention the melting of ice flows or glacier fields on land as the water vapor in most cases will evaporate and become new clouds and never see the ocean. I guess it would be insignificant relative to the level of the ocean,*

as all rivers, streams, rills, creeks, lakes, ponds, and seas also end up in the ocean yet the ocean remains at normal levels as input and output to it remain in balance.

Stott-In one of my comments, I pointed out that land-based ice flows that go into the ocean definitely raise the water level – not when they melt, but when they first enter the water and displace the 7/8th of their volume that sinks below the exposed part of the iceberg.

I also made a statement pointing out the fallacy of the claim that water vapor rising from ice flows or glacier fields on land will never see the ocean.

Stott-However, Schramek's argument does hold water, pun intended. Though he didn't point this out, I will. Most of the ice in the Northern Ice Cap is floating on the Arctic Ocean. Some of the ice is on Greenland and parts of northern Canada, Russia, and Alaska. However, more than half is floating and therefore, if the ice cap melted, only the lesser part that is on land would impact ocean levels. The Southern Ice Cap is different; most of the ice is over land. However, a large part is floating on the Weddell Sea, the Bellingshausen Sea, and the Ross Sea. Taking both ice caps together, roughly half of the ice caps is floating and therefore cannot impact sea levels. I have not heard Gore or any of his friends clarify this and judging by the dishonesty of many of their other conclusions, I am guessing they have not taken this into account. So, rising sea levels – may not be such a big deal.

Joseph E. Schramek

Schramek-*I'm shocked! I'm right!*

Stott: After doing more research, on Sep 21, 2015, I updated the review I posted on my GoodReads page and you quote above as follows: Update: I have checked several Climate Change articles on rising sea levels. In all of them, the rising sea levels are tied only to melting ice on land, namely Greenland, West Antarctica, and ice caps on high mountains. Though the discussion in the book is interesting on some levels, it fails to add anything to the Climate Change discussion. I have reduced my rating from three stars to none.

Stott-This book is a very unconvincing argument against man-caused global warming. That is not to say that extremely convincing arguments don't exist – they do. Schramek just hasn't used them in this book.

Schramek-*It's unconvincing as I didn't write the book for man-caused global warming. It doesn't exist as it requires an increase in energy output from the Sun or a slowing of the rotation of the earth about the axis would lengthen the earth's daily exposure to the sun and it would increase the earth's average temperatures or global warming. My other concern is that the use of wind-mills to create electrical power could exert a reaction force opposite to the rotation of the earth and eventually begin to lessen the speed of rotation, and cause the planet to get hotter.*

Stott-There is well documented and proven history of Ice Ages that are interspersed with interglacial periods

(periods of significant global warming). These periods have not been caused by increased sun radiation nor variations of the speed of rotation of the earth. In addition, there is credible evidence that we're are in an interglacial period that started about 20,000 years ago, so global warming is occurring right now. The question; is man responsible? You and I both say no. The title or your book is "Fallacy of …Climate Change: … "An Inconvenient Truth" By Al Gore." Your title and much of the data and arguments in the book indicate that the book is written to a great extent to argue against man-caused global warming.

Author's Rebuttal to Glen R. Stott's Criticism As Shown in His E-Mail Back to Schramek-August 31, 2018

Glen R. Stott's criticism of my rebuttal to his criticism of My Book in Goodreads Reader's will be shown in regular text and preceded by the name, **Stott**, and my rebuttal to his criticism will be shown in ***bold-faced italicized text*** and preceded by the name, **Schramek**.

Responses to the issues you raised on my comments relating to your issues with them. I'll make my responses for the issues in each paragraph you wrote.

1st Paragraph:

Schramek-"***Extreme weather events are caused by low pressure fronts, 26-27 Barometric Pressure, and not global warming***".

Stott-I live in California. Each year meteorologists report on the water temperatures in the Pacific Ocean. An El Nino condition indicates relatively high temperatures and leads to predictions of a stormy, wet winter in California. A La Nina condition indicates cooler water temps leading to a dry winter. If a global warming condition did exist, warmer waters in the Pacific (El Nino conditions) would become more common and more intense over time, creating increased frequency and intensity of storms in California. Though individual storms can be analyzed in terms of weather fronts as you describe, that analysis has no bearing on global warming. Global warming does impact intensity and frequency of weather patterns over time. So, the question becomes; is there global warming?

Schramek-*I didn't include this topic in my book, because it can only occur with more radiant energy output from the sun or a reduction in the earth's rotational speed at the equator (one revolution every 24 hours) to a longer period of time. Since there is no evidence of this occurring, global temperatures will stay within current ranges of variability over time.*

Stott-I can't recall the exact words of my original rebuttal, but the argument is as follows:

Slower rotation lengthens the daylight hours when the Earth absorbs heat from the Sun, but it also lengthens nights when the heat is radiated into space. Slower rates do not increase the percentage of time the Earth spends facing the Sun. The Sun is the major source of heat on Earth and temperatures do fluctuate with fluctuations of energy from the Sun, however, Earth's temperature is significantly impacted by things other than the energy from the sun. If that were not so, Earth's temperature would be blazing hot in the day and freezing cold at night. It is Earth's atmosphere and the greenhouse effect that makes the Earth habitable. The greenhouse gasses trap Sun's energy to keep the planet warm and livable night and day. If the greenhouse effect is materially changed, Earth's temperature will materially change. That is the basic argument of Climate Change advocates, and your rebuttal above doesn't address that.

Stott-Why is the water vapor molecule responsible for all climatic disasters? I don't recall asking this question or why, since I know the answer; I would need to see the context of this question to decide why I asked it. I agree with your answer. However, the concentration and energy in a mass of water vapor molecules is impacted by heat in the atmosphere, and the amount of heat is impacted by greenhouse gasses. Climate Change advocates argue that increasing CO_2 is causing increased heat in the atmosphere, which impacts concentration of water vapor and energy in water vapor and results in increased extreme weather incidents. If you are going to make a cogent argument against claims of increased extreme weather over time caused by Climate Change, that is the argument you have to address – not water vapor's impact upon individual storms.

Schramek-*Burning of coal and gasoline produces water vapor and carbon dioxide. Both products are greenhouse gases. Water vapor is the more potent of the two greenhouse gases. Water vapor can reach a high of 4% of the atmosphere over the ocean, while carbon dioxide is only .0392% of the atmosphere with an annual range of*

+.0003% in winter months and -.0012% at the end of the growing season. The water vapor molecule has a lighter molecular weight (18) but higher amounts water vapor in the atmosphere (up to 40,000 ppm (parts per million)) for water vapor versus a relatively constant 392 ppm for carbon dioxide-molecular weight (44), and has a much larger impact (100 times) on the earth's atmospheric pressure. Since the amount of carbon dioxide is almost constant at 392 ppm, and the water vapor molecule varies from 0-4% of the atmosphere, water vapor molecule is totally responsible for our violent hurricanes and tornadoes and storm damages from high-straight line winds in storms.

Stott: The conclusion you draw is based upon your assertion that CO_2 is not rising. In other parts, you claim scientific measurements showing rising CO_2 at locations near volcanoes give inaccurate results. What you fail to address is that Climate Change scientists use sophisticated computer simulations to adjust their readings to account for volcanic activities and that there are many other measuring stations that are not near volcanoes which also show rising CO_2.

Schramek-*In high pressure systems (Barometric reading of 29-30 inches of Mercury), we experience good weather with few clouds and fresh air with the highest percentage of oxygen-21% and lowest humidity in the air for comfortable breathing. In low atmospheric pressure systems (Barometric reading of 26-28 inches of Mercury) the air is very warm and humid-3% water vapor and with* *reduced oxygen-about 20% in the atmosphere for poorer breathing conditions and discomfort as perspiration condenses on skin without evaporation and cooling of the skin on our body. Low pressure systems are sometimes incorrectly labeled heavy air. Low pressure systems provide the water molecules which produce rain when the system is confronted with a colder temperature high pressure fronts that generally comes from the Northwest in the Northern Hemisphere. These fronts produce the most devastating tornadoes and thunderstorms with high straight-line winds. For some reason, the scientists ignore the water vapor molecule as the cause for all the climatic temperature disasters and global warming (real or unreal or insignificant). Because there is increasing burning of fossil fuels in the world, they incorrectly choose the other greenhouse gas, carbon dioxide as the culprit, even though the percentage of carbon dioxide in the atmosphere is constantly in the range of 392 ppm or .0392%.*

Stott: Climate Change scientists do not ignore water vapor. They make an argument for excluding water vapor as a cause for global warming. You should research that argument and challenge it if you want to make this point.

Schramek-*Why do hurricanes, cyclones and typhoons always spin?*

Anyone who has ever looked at an image of a hurricane knows it spins. Part of this is due to the center of low pressure — the "eye" — at the center of the storm. But it *also has to do with physics. In fact, tropical cyclones — the general name for the storms called typhoons, hurricanes or cyclones in different parts of the world — always spin counterclockwise in the Northern Hemisphere, and spin in the opposite direction in the Southern Hemisphere. The reason is something called the Coriolis effect, or Coriolis force, named for the French mathematician Gaspard-Gustave de Coriolis, who published work on the effect in the 19th century. It works this way: Like a record on a turntable, the earth spins at a different speed at the equator than it does at the North Pole. The same is true of anything that spins or rotates — the outside edge of something (in this case, the equator) always spins faster than the inside edge. If you placed a marble in the center of a flat plate and then tried to push that marble to the edge of the plate, the marble would move in a straight line, as long as the plate was still. But if the plate was spinning, the marble would follow a curved pattern as it traveled from the center to the edge. Winds passing to and from the North and South Poles and the equator are subject to this effect. Imagine if a person were to stand at the North Pole and throw a ball far enough to reach the equator — say, to a person standing in Quito, Ecuador — the ball would not actually reach that person because it would not travel in a straight line. Since the equatorial region is moving faster than the North Pole, the ball would end up to the west of its target — somewhere in the Pacific Ocean, probably.*

Stott-The above is an excellent description of storm spin. It can be an interesting part of your overall book, but it plays no part in a global warming discussion, so I don't know why put it in this section of your book.

2nd and 3rd Paragraph

Schramek-*There is no evidence that global temperatures are increasing. I didn't include this topic in my book, because it can only occur with more radiant energy output from the sun or a reduction in the earth's rotational speed at the equator (one revolution every 24 hours) to a longer period of time. Since there is no evidence of this occurring, global temperatures will stay within current ranges of variability over time.*

Stott-Taken on its face, that statement denies there ever could have been ice ages and the interglacial warming periods in between. If we admit there was ever an ice age followed by an interglacial, the only two conditions you say could cause it would force us to conclude that the interglacial could only occur if Earth's rotation decreased for a few thousand years or the sun significantly increased radiant energy output for a similar time period. Either contention requires some serious scientific support, which to my knowledge doesn't exist. Without some reasonable explanation, most readers (who are aware of ice age cycles) are going to mark this work as not very serious or reliable.

Of the energy that warms the earth, the percentage coming from Solar radiation is the high nineties, which seems to support your contention taht Earth's temperature should "stay within current ranges of variability over time."

However, Earth's orbit around the sun varies considerably (moving closer to the sun and then expanding away) on a cycle of about 100,000 years. The tilt of the earth on its axis varies on another cycle. These events have significant impact on the intensity of the Sun's radiant energy that hits the Earth over time even though the Sun's output remains relatively constant. These results can have significant impact on Earth's climate and may be responsible, to some extent, for the ice age / interglacial cycles. Over the past 400,000 years, the earth has experienced four such cycles – not caused by man. At this particular time, we are in an interglacial period which is still on the upswing. So, there is global warming occurring. The Climate Change proponents dismiss these facts with rather weak, specious arguments when forced to face them, but mostly, they ignore them because they want to blame global warming on man's activity. In addition, their climate models tend to exaggerate the warming that is occurring.

Schramek-*An interglacial period (or alternatively interglacial, interglaciation) is a geological interval of warmer global average temperature lasting thousands of years that separates consecutive glacial periods within an ice age. I had to look up the word interglacial periods as I had not studied this before. This a common way to add support for the current emphasis for global warming. Some Scientist's emphasize events with great reliance with respect to accuracy on events measured from 400,000 years ago. To me, this is a way to convinced amateurs of their findings. Here are some of my findings and use of facts relating to these matters:*

Joseph E. Schramek

Stott: Climate Change advocates generally ignore interglacial periods because the such periods dilute their claim that man causes global warming. There are several assumptions and large variations that come from using gasses trapped in icebergs as proxies to determine ancient temperatures. When those things are taken into account, the resulting data is very useful in certain climatology studies. Climate Change scientists used to misuse that data, but having been called on it, they now tend to ignore it.

The existence of interglacial periods rebuts your argument that only a change in Earth's rotation rate or in radiant energy from the sun can cause global warming.

Schramek-*There is no evidence that global temperatures are increasing.*

Stott-There is fairly strong evidence that global warming is occurring now and that it started over 20,000 years ago and is still on-going – long before industrialization. Climate Change advocates tend to ignore this.

Schramek-*I didn't include this topic in my book, because it can only occur with more radiant energy output from the sun or a reduction in the earth's rotational speed at the equator (one revolution every 24 hours) to a longer period of time. Since there is no evidence of this occurring, global temperatures will stay within current ranges of variability over time.*

Stott-the cause of interglacial periods is not positively known at this time, but their existence definitely proves global waring can occur without increased energy from the sun, - unless.... see my argument in the paragraph below from **Stott**:

Schramek-*Continued:*

- *The closest the sun gets to the earth occurs on January 3 when the distance to the sun is closest at 147 million miles and because of the axial tilt of the earth, the maximum intensity of the sun at 23.4 degrees above the equator. The Northern Hemisphere is experiencing Winter, and the Southern Hemisphere is experiencing Summer. On July 3, the earth is 152 million miles from the Sun, the Southern Hemisphere is experiencing the maximum intensity of the sun at 23.4 degrees below the equator.*
- **Stott-**This is the current shape of Earth's orbit. This varies significantly in cycles over periods of thousands of years, impacting the amount of solar energy hitting the Earth over time. It is possible that this may play an indirect role in the interglacial phenomenon.
- *Environmentalist's feel that rain water mixes with Carbon Dioxide to form carbonic acid. Keep in mind that the ratio of Carbonic Acid to Carbon Dioxide is .0017 in pure water and .0012 in seawater. Hence the majority of carbon dioxide is not converted into Carbonic Acid when Carbon Dioxide is dissolved in water. The current pH level of oceans is 8.104 is a base and not acidic. Since most of the Carbon Dioxide*

> *mixed with water in rain is not acid but bubbles of Carbon Dioxide. Like a buffer, it makes sense that the Carbon Dioxide bubbles would rise out of the water due to turbulence and re-enter the atmosphere again and even dilute most of the Carbonic Acid until more rain water falls in the ocean. It seems to me that the Environmental Scientists who predict doom and gloom for the ocean and it's marine organisms from projected increases of acidic hydrogen ions from carbonic acid in the oceans as the rise of Carbon Dioxide gas in the atmosphere increases to a pH level of sea water of 7.949 for 2050 and 7.824 in 2100. Keep in mind these values are non-acidic but base or alkaline water.*
- **Stott-**This is interesting, overly simplistic, and not pertinent to the global warming / greenhouse issue we were discussing.

Schramek-*Still the hysteria continues relative to Climate Change persists:*

- *The US Navy expects a 15 cm rise in ocean levels for 1950 to 2000 and the Navy feels this is a top concern for them. I call this no change.*
- *President Obama going to a small island in the Pacific where the island had high water levels and he declares it this his top priority item to fix. Unbelievable!!*
- *All low-lying areas near oceans experience these events when the water is at high tides (moon and sun on the same side of earth) and the winds experiences high speed wind blows from weather fronts coming in from the ocean into the land there.*

- *Never build near the edges of the ocean near sea level. Go to Venice, water covers the side-walks of the business districts under similar conditions.*
- *10 billion metric tons of Carbon Dioxide gets dissolved in the world's ocean each year. Keep in mind, carbon dioxide doesn't affect the acidity of the ocean, the carbonic acid created is only the .12% of total of the 10 billion tons. The ocean is not acidic, it is basic with a pH of 8.069. Actually, the carbon dioxide gas is free to leave the ocean the same it does in a pop or beer. Rain water is a very weak acid of a pH of 5.65. Most will find their way into the stoma of vegetation to produce future energy, building and food products to satisfy our population of inhabitants of the earth.*
- *After 40 years, you can see the little stones in concrete exposed on the top surface of the side-walk in front of my house due to the acid rain.*

4th Paragraph

Stott-Solar energy impacts the Earth just as it does the Moon. However, the consequences on Earth are significantly different than on the Moon. In both cases, the Sun sends energy to the side of the sphere facing it and that energy heats the sphere. On the side of the sphere that faces away from the sun, the surface radiates energy to space and that side cools. A vital difference between the Earth and the Moon is the atmosphere on Earth. This atmosphere contains some gasses, called greenhouse gasses, that tend to blanket the Earth and prevent some of the energy from radiating back into space and at the same time, these gasses shield the Earth from solar energy. The net impact of greenhouse gasses is to keep the planet warm. The more greenhouse gas there is, the warmer the Earth is, in spite of no increase in the inflow of energy from the Sun. As you point out, by far the most prevalent greenhouse gas is H_2O. The amount of CO_2 is nearly insignificant. Climate Change advocates come up with a lot of complicated, sophisticated arguments and analyses to convince us the slight changes in the tiny amount of CO_2 change the climate on Earth. Those arguments must be addressed for a credible argument against Climate Change doctrine.

Schramek-*Added information on the relative amounts for Solar Radiation effects on the Earth and Moon temperatures, CO_2 and H_2O gas and water vapor effects on Earth temperatures, and some more warming of ocean temperatures at the South Pole as shown below:*

- *Earth-Solar Radiation-35% is reflected, 17.5% Scattered to the Earth from blue sky, 10.5% Scattered from clouds, and 22.5 % going to the Earth's surface. The topmost layer of the atmosphere must radiate to space the same amount of energy it received from the Sun. If this were not the case and the topmost layer radiated less, the planet would heat up, or if it radiated more, the planet would cool off. As a consequence, the planet's temperature will change until it radiates as much as it receives.* **Stott:** You have finally hit upon the problem. The crux of the Climate Change argument is that increasing CO_2 (greenhouse gas) causes the top layer to radiate less energy thereby causing global warming. True, most of the CO_2 gas is close to the Earth, however, that gas reduces the amount of heat that reaches the topmost layer, thereby, reducing the heat that radiates from the top. The question to address is, can the small increases in CO_2 gas create enough impact to significantly raise Earth's temperature. Climate Change scientists claim they have convincing evidence that it does. You cannot invalidate that evidence and their interpretation of it by simply claiming it doesn't exist. You need to study the evidence and their interpretation to make an informed rebuttal. That is not straight forward nor easy; if it were, there would be no Climate Change crisis and no calls for global Cap and Trade laws.

- *The earth captures .0000000045% of the Sun's radiant energy output.*
- *Moon-Solar Radiation affects only the surface, as there is no atmosphere. For full moons, the temperature of the surface facing earth at is 253 degrees F or less depending on the fullness of the moon. The opposite of the moon non- illuminated is –243 degrees F. Astronauts must have had special boots to handle the high temperature of the surface.*
- *The argument that a tiny amount of CO_2 will raise the earth's temperature lies in the fact that the CO_2 molecule has a molecular weight of 46 versus the weight of the Water H_2O molecular weight of 18. This makes little sense, since the CO_2 molecule becomes acid rain H_2CO_3 during rain-fall. Being heavier, CO_2 gas stays closer to the ground due to gravitational pull of the earth.* **Stott:** Only a portion of the CO_2 goes into

acid rain, fortunately; because all life would die without free CO_2 in the air.
- ***The switch to Methane or Natural Gas CH_4 for Electrical Power Plants actually increases greenhouse gases as the product of combustion gives 1 molecule of CO_2 and 2 molecules of H_2O Water. Water is a greenhouse gas.*** **Stott:** Methane is a significant greenhouse gas. Climate Change advocates would go after it but, I think, they like their steaks too much and going after meat producers does not give political advocates near the money, power, and control as regulating all energy production on the planet.
- ***Environmentalists Claim a Warmer Climate at the South Pole Causes More Ice to Form. Environmentalists claim that the ice at the South Pole is increasing in warmer weather because the ice (pure water) dilutes the ocean salt water and the ice freezes at a higher temperature.*** **Stott:** The hoops they must jump are interesting. They had to change the name of their movement from Global Warming to Climate Change to avoid raised eyebrows when they said cold spells were caused by Global Warming.
- ***Any ice that floats off land will never increase the water level of the ocean if melted as the ice is .9 the density of water where 90% of the ice berg is in water and 10% is above water.*** **Stott:** Any time an iceberg moves off land into the ocean the 90% of it that sinks into the water must displace the water. That displaced water raises the level of the ocean when the iceberg enters the water. True, melting won't raise the level any more, but the damage has already been done. ***Any ice melting of ice on land***

that flows into the ocean is just like the rivers of the world that flow into the ocean and any water added to the ocean will increase the circumference of the ocean but daily evaporation of the ocean's water into the atmosphere has equalized the levels of the oceans forever. **Stott:** What an incredible claim! Forever? The rate of evaporation of water from the oceans is not in the least bit impacted by the flowrate of water from rivers or other sources, and since they are not related, they cannot get into equilibrium. Sea level is somewhat in balance now, but if the inflow of water is increased dramatically by melting ice from the poles and high mountain tops, the sea levels must rise to accommodate the increased amount of water in the ocean. On the other hand, if an ice age reduced water flow into the ocean by freezing it, the oceans would have to drop. *The earth can't get bigger because it is stuck with the same number of atoms over time with the exception of Hydrogen gas. Few Hydrogen gas molecules are the only ones that can escape the earth's gravitational pull. Something I learned in college. The Hydrogen molecule probably goes back to the Sun. Because of the intense gravitational pull of the earth, I believe we are stuck with the same weight, roughly the same number atoms and the current size of the earth.* **Stott:** Increased water in the oceans and rising sea levels may increase the circumference of the ocean, but it has no impact on the circumference (size) of the Earth. The circumference of the ocean is not the circumference of the Earth and, in fact, the circumference of the Earth is significantly above sea level. It is an average, including land and sea. If all the land and ice melted and flowed into the sea and only water was visible all around the Earth, then sea level would be the circumference of the Earth and would not vary, but no water would be flowing in either. The fact is, since ice occupies more volume than an equal weight of water, all the melting ice would result in a slight diminishing of the circumference of the Earth.

Since these matters may take years to substantiate the issue or to show the issue has no merit to the degree the sponsor's are predicting, I feel comfortable with my position. Carbon Dioxide is the miracle gas that that enables human, animal, and marine organisms to survive on earth forever until the sun is extinguished. It provides us oxygen, food, vegetation, and energy products via photosynthesis and energy from our Sun.

Finally, the "Fallacy of Climate Change and the Green Movement" may come to an end, as on May 5, 2018, the NASA National Atmospheric and Space Administration began bragging that their station on Mauna Loa report the CO_2 levels increased to 415 PPM. The very next day, the Kilauea volcano erupted and lava flow filled the South End of Hawaii Island and CO_2 filled the island and recording site at NASA. My original book use the argument that the NASA site was in error for measuring CO_2 at 11,000 feet near vents and crevices of volcano sites and this gave false results on CO_2.

Schramek-*I want to thank Glen R. Stott for providing his rebuttal criticism and supporting comments via Goodreads Reader's on September 21, 2015 and again his rebuttal responses to my rebuttal comments via E-Mail on August, 21, 2018 on some of my positons, and his approval to use these positions in my reissue of my book, "Fallacy of the Green Movement and Climate Change" published by Xlibris.*

Foods High in Lead Content

Lead content is in mcg per 100 g of food weight.

Lead	Food Name
26.5	Mussel, green, steamed or boiled
16.5	Lamb, kidney, simmered
16.5	Lamb, kidney, simmered
11.4	Bassa (basa), steamed or poached
9.8	Barramundi, aquacultured fillets, steamed or poached
9.7	Bassa (basa), fillet, raw
8.6	Silver perch, aquacultured, steamed or poached
8.3	Barramundi, aquacultured, fillets, raw
7.3	Silver perch, aquacultured, raw
7.0	Salmon, Atlantic, fillets, raw
6.5	Milkfish, aquacultured, steamed or poached
6.2	Trout, rainbow, aquacultured, steamed or poached
6.0	Venison, mince, premium, dry fried
5.5	Milkfish, aquacultured, raw
5.2	Trout, rainbow, aquacultured, raw
4.7	Lamb, liver, grilled
4.2	Venison, mince, premium, raw
3.8	Sultana
3.6	Chocolate, milk, with dried fruit & nut
3.1	Dried fruit mix (raisin or sultana), milk chocolate-coated
2.9	Crabmeat, canned in brine, drained
2.8	Venison, stir fry strips, lean, dry fried
2.6	Crabmeat, canned in brine

2.4 Tilapia, steamed or poached
2.2 Chocolate, milk, with added milk solids
2.0 Tilapia, fillet, raw
2.0 Biscuit, sweet, plain
2.0 Dried fruit & nut mix, milk chocolate-coated
2.0 Potato, coliban, peeled, boiled
1.9 Venison, stir fry strips, lean, raw
1.8 Wine, white, medium sweet style (~ 2.5% sugars)
1.8 Wine, white, medium dry style (~ 1% sugars)
1.8 Wine, white, dry style (sugars content < 1%)
1.7 Tahini, sesame seed pulp
1.6 Cabbage, bok choy, stir-fried without oil
1.6 Bread, from wholemeal flour, toasted
1.6 Cheese, feta (fetta), sheep & cows milk
1.6 Seed, sesame, white
1.4 Bread, from wholemeal flour
1.2 Pizza, supreme, purchased frozen, baked

Lead Food Name
1.2 Pizza, supreme, thick base, fast food chain-style
1.2 Pizza, supreme, thin base, fast food chain style
1.2 Pizza, ham & pineapple, purchased frozen, baked
1.2 Pizza, ham & pineapple, thin base, fast food chain-style
1.2 Pizza, supreme, takeaway style
1.2 Pizza, ham & pineapple, takeaway style
1.2 Pizza, ham & pineapple, thick base, fast food chain-style
1.1 Cabbage, bok choy, raw

1.1 Biscuit, savoury, cheese-flavored
1.1 Potato, red skin, peeled, boiled
1.0 Breakfast cereal, mixed grain (wheat, oat & corn), extruded shapes, added vitamins B1, B2, B3, B6 & C, Ca & Fe
1.0 Biscuit, savoury, rice cracker
1.0 Breakfast cereal, wheat bran, flakes, sultanas, added vitamins B1, B2, B3, B6 & folate, Fe & Zn
0.9 Breakfast cereal, mixed grain flakes (wheat, oats), added dried fruit, added vitamins B1, B2, B3 & folate & Fe
0.9 Almond, milk chocolate-coated
0.9 Prawn, king (large size), flesh only, purchased cooked
0.9 Prawn, school, flesh only, purchased cooked
0.8 Salami, danish
0.8 Salami, unspecified variety
0.8 Salami, mettwurst
0.8 Salami, pepperoni
0.8 Lettuce, cos, raw
0.8 Salami, hungarian
0.8 Lettuce, mignonette, raw
0.8 Salami, milano
0.8 Lettuce, iceberg, raw
0.7 Avocado, raw
0.6 Sausage, beef, fried
0.6 Cake, chocolate, iced, commercial
0.6 Nut, walnut, raw
0.6 Potato, pale skin, peeled, boiled

0.6 Hamburger, plain (beef pattie, lettuce, tomato, onion, sauce), takeaway shop
0.6 Macadamia, milk chocolate-coated
0.6 Water, tap
0.6 Peanut, milk chocolate-coated
0.6 Sausage, beef, grilled
0.5 Soy beverage, reduced fat (~1% fat), unflavored, added Ca & vitamins A, B1, B2 & B12
0.5 Sprout, alfalfa, raw
0.5 Cucumber, lebanese, unpeeled, raw

Lead Food Name
0.5 Cucumber, telegraph, unpeeled, raw
0.5 Soy beverage, reduced fat (~ 1.5%), coffee flavored, added Ca & vitamins A, B1, B2 & B12
0.5 Soy beverage, regular fat (~3%), unflavored, added calcium
0.5 Soy beverage, regular fat (~3%), chocolate flavored, added Ca & vitamins A, B1, B2 & B12
0.5 Breakfast cereal, whole wheat, flakes, added dried fruit & nuts, added vitamins B1, B2, B3 & folate, Ca, Fe & fibre
0.5 Milk, cow, fluid, reduced fat (1.5%), increased Ca, folate & vitamin D
0.5 Soy beverage, reduced fat (~ 1.5%), chocolate flavored, added Ca & vitamins A, B1, B2 & B12

0.5 Pasta, white wheat flour, boiled from dry, no added salt
0.5 Soy beverage, regular fat (~ 3%), unflavored, added Ca, vitamins A, B1, B2 & B12
0.5 Bread, from white flour, toasted
0.5 Egg, chicken, whole, omega-3 polyunsaturate enriched, raw
0.5 Cucumber, common, unpeeled, raw
0.5 Bread, mixed grain, toasted
0.5 Soy beverage, regular fat (~3%), unflavored, unfortified
0.5 Breakfast cereal, beverage, all flavors, added vitamins A, B1, B2, C & folate
0.5 Egg, chicken, whole, omega-3 polyunsaturate enriched, boiled
0.5 Milk, cow, fluid, regular fat (3.5%), added omega 3 polyunsaturates
0.5 Bread, from white flour, added omega-3 polyunsaturates, toasted
0.5 Cheese, cheddar, regular fat
0.5 Cheese, cheddar, processed, reduced fat (~16%), added vitamin D
0.5 Pasta, white wheat flour, boiled from dry, with added salt
0.5 Chicken, breast, lean, grilled
0.5 Milk, cow, fluid, reduced fat (1.5%), added omega 3 polyunsaturates
0.5 Cucumber, apple crystal, unpeeled, raw
0.5 Cheese, cheddar, processed, reduced fat (~8%), added vitamin D
0.5 Mushroom, common, stir-fried without oil

0.4 Venison, leg medallion, lean, dry fried
0.4 Venison, leg medallion, lean, raw
0.4 Bread, mixed grain
0.4 Cake, lamington, plain, commercial
0.4 Bread, from white flour, added omega-3 polyunsaturates
0.4 Beverage base, chocolate flavor, added minerals calcium & iron & vitamins A, B1, B2, C & D
0.4 Bread, from white flour
0.4 Venison, diced, lean, dry fried
0.3 Chicken, drumstick, lean, baked without oil
0.3 Chiko roll, deep fried
0.3 Venison, diced, lean, raw
0.3 Beetroot, canned, drained
0.3 Tea, regular, brewed from leaf or teabags, without milk

Epilogue

My hope is that 25-50 years from now, someone or a grandson or grand-daughter or a relative or a friend of the family will find this book at a Salvation Army discount books stand, or a garage sale. Hopefully the finder will have enough knowledge of Chemistry, Physics, Electricity, and Math to fully comprehend the factual message in each paper included in the book that relates to the atmospheric conditions or electrical generation covered in the individual papers. Hopefully, the finder will make sure that I'm identified as the first person to accurately tell the truth about the fallacy of the Green Movement and the Big Lie About the Inconvenient Truth by Al Gore. By the time the book is found, others will come to the same position on this issue as stated in this book, but this book makes me the first to accurately show the fallacy of their arguments.

I wrote this book and its articles when it became clear that both sides of the Climate Change, or Green Movement, or Global Warming alarmists had no knowledge of the elementary Chemistry, Physics, Math, and Electrical Power generation involved in how our atmosphere makes weather events that cause beautiful days, cloudy days, cold days, rainy days, snowy days, tornadoes, cyclones, straight-line winds, and hurricanes. All of these events are normal occurrences of the atmosphere. We learned as children in the story of the 3 Little Pigs, that those who make their homes of straw, and wood, had their homes destroyed by the Big Bad Wolf who was able to blow them apart. Only the pig who made his home of bricks survived the attack by the Big Bad Wolf. He was unable to blow the bricks apart and the Pig lived as a result. Very few people have learned this lesson, and some of those paid their price when they were hurt or killed when a major destructive wind storm hit their home. No one wrote a book about making your home along a stream, river, rill, lake, sea, or ocean would be like making your home out of straw or wood as told in the 3 Little Pigs tale.

I feel sorry that President Obama, Environmental Protection Agency, most of the democrats and a few republicans, environmentalists and their so-called Scientists, Al Gore, pseudo stewards of the earth including the Pope and US Bishops and Nuns and others, and those in some media outlets and press that full heartily supported the Green Movements positions on everything, because they were lacking the education in Chemistry, Physics, Electricity, and Math and interest in the function of our atmosphere in the growth of vegetation and oxygen generation that sustains life here on earth.

Since the universe is estimated to be 13.7 billion years old, and the fact that all the atoms of each of the chemical element will remain here on earth until the universe and the earth collapse, except for those included in spacecrafts sent into deep space never to return, and hydrogen gas molecules that develop enough velocity to escape the earth's gravitational force. Only the hydrogen gas molecule can develop enough speed to leave the earth's gravitational pull of the earth. The hydrogen molecule is rare and by volume is .000055% of the total atmosphere.

On a personal note, I always asked my dad a lot of questions about things on how, why, when, or where about things I couldn't understand. The classic question was how do you spell a particular word and he always said look it up in the dictionary. How can I, I don't know how to spell it. My mother fixed that problem when my marks were drifting South, she made me read the dictionary one summer. My son, Joe, wore a shirt that said I survived 12 years of Catholic School Education. I felt the same way.

I may not be able to convince some people about the Fallacy of the Green Movement, but they may be able to use some of the information here on the atmosphere.

www.ingramcontent.com/pod-product-compliance
Lightning Source LLC
Chambersburg PA
CBHW030942180526
45163CB00002B/677